Dear Bruce,

It's been great working with you these last few years.

I trust you will enjoy reading about Jas and his many adventures while practicing industrial hygiene globally.

Like Jas my mentor and friend, I encourage you to make your IH career an amazing one!

Shanini
March 2014

Jas

Jas

Chronicles of Intrigue, Folly, and Laughter
in the Global Workplace

Jas Singh, PhD

with

Gregory Beckstrom

TWO HARBORS PRESS

Two Harbors Press
322 First Avenue N, 5th floor
Minneapolis, MN 55401
612.455.2293
www.TwoHarborsPress.com

ISBN-13: 978-1-62652-551-1
LCCN: 2013922007

Distributed by Itasca Books

Cover Design by Carol Nagan
Typeset by Kristeen Ott

Printed in the United States of America

Table of Contents

Foreword

It is a great honor and pleasure for me to write the foreword for the book by my friend, past colleague, and peer, Dr. Jas Singh. Jas has spent his entire career in industrial hygiene and environmental health and has a long track record of accomplishments, including: President of the Academy of Industrial Hygiene (AIH), Director for the American Industrial Hygiene Association (AIHA), Director for the American Board of Industrial Hygiene (ABIH), and Editorial Board member of Applied Occupational & Environmental Hygiene. He is the winner of the Distinguished Service Award from AIHA and the Henry Smyth Award from the AIH. With all of these accolades and a PhD in physical chemistry, one would think of Jas as a highly qualified professional but probably quite boring. Nothing could be further from the truth. "Jazz Man" (as I like to call him) is one of the most extroverted, adventuresome, and fun persons I know.

They say it is a small world, and for Jas and I, this has been proven time and again. Early in my career, I certainly had heard of the renowned Dr. Singh as he worked for one of the largest and most active industrial hygiene consulting firms of the day, Clayton Environmental Consultants (now part of Bureau Veritas). In fact, Jas was a leader in that organization. He was not just a bureaucrat but someone who found (and still finds) excitement in field work where he can deduce the cause of a problem. I first met Jas formally when I worked at Midwest Research Institute in Kansas City, Missouri. We were part of a consulting team that was assembled to tackle a very large nation-wide project for the defense department that was being managed by Bolt, Beranek & Newman (BBN is now part of Raytheon) out of Boston. This was circa 1980. I found him full of technical knowledge, as one would expect, but also very much the charmer and with quite a sense of humor and adventure. Little has changed in the man I first met more than thirty years ago, though I hope he no longer wears bell-bottom suits. Our lives have certainly been very connected as friends, peers, and coworkers. If we can fast forward to the early 1990's, I had established a boutique environmental consulting firm, Mansdorf & Associates, Inc., in northern Ohio which was being courted by Clayton Environmental Consultants and was ultimately purchased in 1994 by Clayton. My new boss was their senior vice president, Jas Singh. Even though I was not thrilled that I had to stay another year to assist with the transition of the business, I was delighted to work with Jas. While I worked with him as a colleague at Clayton, I found Jas to be extremely professional, very well respected, and a fun travel companion. He was also a very good businessman. Jas and I got to know more and more about each other, which would prove fruitful in the future.

Fast forward a few years ahead again. I was hired by Liberty Mutual Insurance Company to head a new company called Liberty International Risk Services. My job was to establish a reputation for Liberty International as a leading provider of expert advice and services related to loss control and insurance outside of the USA. As part of the job, we wanted to open a new business in India. So I called Jas. He was the very best person in the world to do what we wanted. He had all the technical and business skills, as well as being a super salesman who could court new clients. I also knew that Jas was ready to seek another challenge. Most persons would fret about being a stranger in a strange land. Jas thrived on it. As fate would turn out, Jas ended up going to Kula Lumpur, Malaysia, instead of India, to run an existing business for Liberty International. The fact that he was there for five years and recruited and grew a capable staff and the business is a testament to his sense of mentorship and adventure. Meanwhile, I went off to Paris for L'Oreal as their head of EHS, and Jas moved his home base in Hawaii to be closer to Asia (another testament to itchy feet). It was not long after that I received a note from him telling me of his work with Golder Associates, who had acquired several of the Liberty International overseas consulting offices. Jas traveled the world for them, and I have the proof from a picture of Jas and a mango tree that I planted at a new factory dedication in Pune, India (actually, the first picture he sent was of a sickly tree that had not grown at all followed later by the real picture with a very healthy tree bearing mangos—another demonstration of his sense of humor).

Those who know the Jazz Man know him as an expert storyteller. I am confident that you will enjoy his adventures around the world, his cast of characters, and his skill at weaving the stories. I hope you like this book as much as I have.

Thank you, my friend, for all you have done for me and for our profession.

Zack Mansdorf, PhD, CIH, CSP, QEP
Boca Raton, Florida

Acknowledgments

I could not have completed this project without the daily assistance of two individuals who, despite being busier than me, adopted this project as their own. First I want to thank Mary Singh, my wife and partner of forty years. Mary is one of the best English editors, who put aside millions of chores to help get my stories written. She also helped me recall many details of stories she had been part of herself and remembered more names and places than my feeble memory allowed. She would often run to the nearby Starbucks® coffee shop with her computer because our Wi-Fi was frequently down, due to the twenty-five-mile-per-hour Hawaiian trade winds. Nonetheless, she always met every deadline.

Second, I thank my friend and partner in crime, Gregory Beckstrom, who toiled as my chief editor, researcher, wordsmith, and fact checker. He also provided direction for all of the maps, illustrations, and photos that you will see throughout the book. Greg and I worked together for several years at Golder Associates, a global ground engineering and environmental services company, and shared some of these adventures. In addition to being a top-notch technical professional (unlike me, he is a geologist), Greg is also a father, science writer, editor, business manager, and global traveler who possesses the same passion for life ("life is to be experienced") as I do and a similar irreverence for many things that others take far too seriously.

Greg also introduced me to Carol Nagan, a talented artist and a wonderful human being. As our illustrator, Carol took a greater interest in this project than just another freelance gig. She is a true associate in this endeavor, or shall we say folly. A great example of this is Carol's rendition of "Jeeto"—the rustic Indian girl who looks like an authentic Punjabi village girl. Although Carol has not visited India, she brought Jeeto to life, as she did many of the other characters and situations throughout this book.

Thanks also go to my daughter Monica and my son David, who were the inspiration for some of my early stories. Their curiosity about camels helped me realize what life-changing events caused me to become a chemist and safety man. They also provided useful critiques of my stories. Their views and comments were always refreshing and honest ("Dad, that would be a dumb thing to say") because they know their dad the way all kids should, in my opinion.

This book is possible due to the encouragement and participation of many others in the environmental, safety, and health/industrial hygiene field to whom I sent my stories for review and/or who gave me their permission to share our mutual experiences with the world-at-large. While we attempted to protect some people's identities, they will know who they are when they read the book. The list of real people is long and includes: Dr. Zack Mansdorf, retired vice president of L'Oreal Corporation; John Henshaw, past US OSHA Director; Brian Senefelder, senior vice president, Golder Associates; Shamini Samuel, manager of health and safety, Suncor Corporation, Calgary; Dr. Dietrich Weyel, professor and consultant, Pittsburgh; Doris Wunsch, EHS Consultant; Mr. Bakini Nor, entrepreneur, Kuala Lumpur, Malaysia; Kim Heron, EHS consultant, Calgary; Chitra Murali, Chennai, India; Andreia Miguel and Anna Paula Medeiros, Sao Paulo, Brazil; William Zhu, Shanghai, China; Bill Eissler, Detroit; "Dancing Debbie" Dietrich, SKC Corporation, Pittsburgh; Charles Blake, Atlanta; Barbara Dawson, DuPont Corporation, Delaware; Tom Barnett, Ministry of the Environment, Ontario; Dan Macleod, ergonomist, New York; Chris Lacz Davis, San Francisco; Robert

P. Allen, Huntsville, Alabama; Norhazlina Mydin, PETRONAS, Malaysia; Rene Ramaswamy, TV show producer, Minneapolis; Robert Sheriff, New Jersey; and Cathy Liu, Singapore. I also thank Coleman Hahn, Texas; Toxic Joe, New Jersey; Svetlana, Moscow, Russia; and Uncle Azad and his beautiful nieces in Azerbaijan. And I also want to acknowledge Cristalina, wherever she is, for welcoming me to the United States of America when I had only a couple coins in my pockets.

In addition, I also owe a debt of gratitude to the late Chain Robbins of US Steel Corporation who blazed paths for many of us to follow.

Jas Singh, PhD, CIH
Kamuela, Hawaii
December, 2013

CHAPTER 1

JEETO – THE UNCUT DIAMOND

This is not a story about industrial hygiene, air pollution, or chemical poisons
It is about dreams and growing up
It is a true story
It is an unhappy story

As a teenager in India I was addicted to "Bollywood" movies. These movies – which are unique to India – were designed to fulfill all your fantasies: palatial homes, beautiful women, exotic food, action, dancing, and music and more music even when the moment did not call for music.

The formula for success was standardized. You could not hope to make a commercially viable Indian film unless you had all the standard ingredients. The recipe had been perfected from hundreds of trials: boy meets girl and immediately sets out on a complex but predictable romantic journey that starts with the boy teasing the girl by singing amorous songs. Such unwanted displays of intimacy would at first irritate and offend any respectable woman. At first she would strongly resent such crude attention and tell the pursuer to get lost. She would have nothing to do with the likes of him. But wait. This tongue lashing would be temporary. Soon the stone would start softening as the suitor persisted. Softening of the stone would be evident when the same tongue lashing was delivered in the form of a song replete with full orchestra in the background even if the encounter was at a lonely public park. The musical rebuff would be like the song from my favorite Bollywood movie called "Mr. and Mrs. 55" (1955).

1

और कोई घर देखिये

(Go look for some other venue.)

दिल को यहाँ मत फेंकिये

(Don't throw away/waste your heart here.)

The heroine singing the song would be the most gorgeous girl around and would be dressed to the hilt even if the venue was a dilapidated city park. If the park happened to have a few decent trees, the boy and girl had to run around the trees several times, and that would often blossom into a game of hide and seek. (Bollywood movies have come a long way since my youth. Although the basic theme remains the same, the boy and girl are now flown to places like Switzerland to run around more lush trees on the Swiss mountains or roll over in piles of fresh white snow.) The boy then would break into his own song and plead for permission to peek inside the girl's *aanchal*, which is the long head-to-toe wrap customarily worn by Indian women. Sometimes he would be more daring and ask the girl if he could hide under her aanchal (*mujhe aanchal mein chhupa lo*), knowing full well the aanchal did not have enough space to hide and moreover was made of a transparent fabric that prying eyes could see through. If the hide and seek went as expected, then the path from there proceeded smoothly, punctuated only by temporary setbacks that the viewer knew would be overcome without a serious glitch.

As a kid I used to wonder where in the world such women were found. They looked Indian but they were nothing like the babes (I had not discovered the term at that time) in my village. I knew this because if there were women like those in my village, I would have noticed them.

Not finding the kind of women, the palatial homes, the night clubs, and the glitter in my village propelled me to the movies every weekend to experience such luxury. I had become a movie fanatic despite my mother's objections. She wanted me to study algebra.

As a teenager, I did not live in the village of my birth. I lived in a larger city (if you could call it that) in my State where there was a middle school because my birth village did not offer education beyond the fourth grade. I would return to my birth village only during summer holidays and for Diwali (the festival of lights) and for an occasional wedding or funeral.

My birth village had no facilities or opportunities for amusement, but I still liked it there. I liked to go to the fields with my cousin, who never went to a school and died an uneducated farmer. He had, however, a fantastic sense of humor. For some strange reason, I also liked the unpolished and unsophisticated girls in the village despite the fact that most of them seemed perpetually covered with a fine tropical dust and a top thinner layer that looked like organic microfibers composed mostly of cow dung.

The cow dung pyramids

Cow dung reigned supreme in my birth village. Heaps of the fresh, green aromatic byproduct, which was a valuable commodity, waited to be transported to the open yards either to dry out for later use as fuel or to be composted. This was truly the green revolution—one hundred percent natural, organic, and biodegradable. Larger quantities, destined for composting, were transported by machines and loaded by burly mustachioed men with muscles hardened and dried under the hot North Indian sun. Smaller quantities of the green gold were converted into perfectly shaped discs (cow patties), dried in the sun and stacked into smaller and smaller circles from the base up, forming perfect pyramids rivaling the Egyptian masterpieces but on a mini scale. The architecturally symmetrical structures were later coated with layers of mud and clay to insulate them so tightly that even a molecule of water could not penetrate. After the relentless monsoons, when everything else was soaking wet, you could open these gigantic hives and find bone dry organic fuel that contained no trace of manmade additives and toxins. Breaching the pyramids to recover dry fuel in the cold and wet season was like finding Tutankhamen's treasure. The first time I witnessed the breaching of one of the monstrous pyramids I felt like Howard Carter, discoverer of King Tut's tomb.

Much of the hauling of the green gold was carried out by village girls between the ages of twelve and eighteen. They would haul the semisolid material that had the consistency of creamed spinach for long distances to a common work area. There they would turn the precious commodity into discs by forming round patties with their bare hands, arranging the discs into neat rows to dry out under the hot sun. They would then neatly shape the dried discs of dung into the symmetrical pyramids. I was certain that their labor sustained the daily life in the village by providing light, heat, and fuel for the cooking pits. You might say they were responsible for fueling the economy of my village.

While the women and girls worked, the men just sat around all day yelling insults at the young boys who were hired to tend to the cow or buffalo herds. Bullsh!t flowed unchecked during such gossip gatherings, but, unlike the cow dung, no one was designated to clean it up.

The Via Candiotti of Gurne

Paris has the Champs Élysées. Rome has the Via Veneto. And Milan boasts of its fashionable Via Condotti where you can shop or simply ogle the beautiful people. But the Via Condotti of Gurne (my birth village) was the street on which my house was located. All the beautiful people went by it. So did the ugly ones. The street had no name. None of the streets in the village had a name or a street sign. To find an address, you had to find someone who would know the person(s) you were seeking. It may sound complicated, but it always worked. It was more reliable than MapQuest or a GPS system. Usually someone who knew the house would escort you, and sometimes not before you accepted a cup of tea at the escort's own house.

On summer days I would sit on my front porch on a jute cot with a big fat pillow to support my back just wasting away time. That was all I did during my summers except for ogling the girls hauling the green gold to the fields to be shaped into symmetrical discs.

On the way to the cow dung disc-forming sites, the girls would walk right in front of my house. My porch afforded me a front-row seat and an unobstructed view of the workers carrying this raw material. No one could wrongly accuse me of staring at the girls carrying the smelly stuff. Staring is generally *not* frowned upon in India. To the contrary, staring at people is a national pastime. It should be known that while staring was permissible, any verbal and/or physical contact was strictly forbidden.

After my first couple of days back home during the summer of my fifteenth year, I started noticing certain patterns. Every day at almost the same time, a pretty girl, maybe fourteen or fifteen years old, would emerge from the right side of my field of vision and soon appear in full glory.

My heart would skip a beat at the sight of her, for even under the sagging weight of probably seventy pounds of dung packed in a leaky cane basket, she radiated beauty, grace, and charm. I was smitten. I described my feelings to my brother's wife, who was my ardent supporter and confidant. Mentioning such feelings to my mother would have meant immediate assignment to other more productive activities, including studies, sports, or some sort of community service, the kind of activities way down on my list of priorities. My mother would have done anything to steer me away from the staring game in which I was engaged.

My sister-in-law Deepa was an angel. She did not approve of my sudden new crush, but she was willing to share with me information about this newly discovered treasure. She told me that the girl's name was Jeeto and that she was fifteen years old. Jeeto lived on the east side of the village where many people of her extended family lived, which was not too far from our house. Deepa, however, warned me not to get too ambitious for several reasons.

Jeeto was underage (as was I at the time). Jeeto was from a different ethnic group, a major obstacle to advancing the scope of my project. Moreover, if my mother found out, I would be banished back to my dormitory in the drab city or advised to visit relatives in another village. Worst of all, if Jeeto's father found out, Jeeto's path would be diverted away from the Via Condotti of Gurne.

None of these alternatives sounded attractive. I assured Deepa that she did not need to worry. I would do nothing stupid—a false promise—to bring shame to our or Jeeto's family.

My observation pad

I was hooked. The next day I made sure I was well positioned on my woven jute cot when Jeeto came by. As if controlled by an atomic clock, Jeeto appeared from the right at precisely the same time as the day before. She noticed my behavior and seemed willing to play along with my game. Jeeto wore the same long Punjabi salwar that she had worn the day before. Sporting an erect posture, even under the seventy-pound weight on her head, she walked as if she were a seasoned model for a Milano fashion house, which, of course, I did not know about at that time. Bracing seventy pounds of cow manure, she still exuded the grace, pride, and confidence of an Italian model. When she approached the same latitude I was, she ever so slightly tilted her head—probably no more than an ergonomically safe 15 degrees— toward me but only for a second or so it seemed. During that flash she gave me a beautiful but subdued smile. Then as quickly as she had tilted her head toward me, she withdrew it in one quick action as if to say: *Okay, this is all you get. This is all you deserve, you slacker. Show me why you deserve more. Earn it if you can.* She then resumed her walk as if nothing had happened.

I slumped in my jute bed. I was mush. I was putty, silly putty at that. I could not resist telling Deepa of my plight, which resulted in another friendly warning.

"Watch out, kid, you are getting into muddy water."

I assured Deepa of no potential harm but continued the cat and mouse game. Every day Jeeto would appear at the same precise moment and look at me. The duration of her glances got longer and the smiles broader. Her eyes were now clearly saying, *Hello, handsome. How are you today? Why aren't you studying hard so you can become a doctor and then ask my parents for my hand?*

I have no proof that Jeeto thought such things. It might have been just my imagination.

I often dreamed of going to America and taking Jeeto with me. It did not matter if she had no college education or could not speak English. I would give her a crash course in English. She would go to school. She would learn American etiquette, and when we got to Los Angeles, I would take her to a fashion house in Beverly Hills to get the diamond polished. Los Angeles was the only city in America I knew anything about. I had heard wonderful things about Hollywood, Beverly Hills, and the Sunset Boulevard from a distant relative who had lived in California.

In my daydreams, when the Beverly Hills fashion house was finished working on Jeeto, she would rival any model in the land of fruits and nuts. I would walk with Jeeto hand-in-hand on Sunset Boulevard and heads would turn. The rich California boys would have their fancy cars and their platinum blonde girlfriends, but only I would have Jeeto. They would stare at me and be filled with jealousy wondering, *where did this farmer guy pick up this diamond?*

No longer could I contain my feelings. I wanted to share my dreams. The next day at breakfast (my favorite meal of the day in my village), I casually blurted: "Mom, don't you think that Jeeto girl is gorgeous? The way she walks and smiles. What strong muscles she must have carrying a ton of that disgusting stuff on her pretty head and yet her beauty and grace shines through from under all that grit."

Disaster control

My mother's jaw dropped. She smelled big trouble, both literally and figuratively. Her baby was growing up. The hormone harvest had arrived prematurely. She reacted as if I had just announced that I had a terminal disease. She immediately went into disaster recovery mode, pulled me close to her chest like when I was five years old. She lowered her voice so no one could hear it even though there was not a soul in sight. She cupped both hands around my chubby cheeks and said: "Listen, son, you have a beautiful future ahead of you. You are so young. One day soon you will be a doctor. People will greet you with folded hands. They will think of you as a life savior. Parents of well-educated girls from prominent families in Patiala or Delhi will be knocking on our creaky door to engage their daughters to my Jas." She gave me a big squeeze, her eyes filled with moisture.

She composed herself and continued: "Just be patient. You will have plenty of girlfriends when it is time. You are such a handsome boy. I know you are destined for America. You could even be the heartthrob of some American lady." She knew the exact Punjabi equivalent of heartthrob, but the word escapes me now.

Then she got to the point. "Why are you wasting your precious time dreaming about this uneducated and unsophisticated village girl with spindly legs and an unstable frame like a baby Shuturmurg (an Indian ostrich)?" Jeeto was tall for her age. "She is always covered with dirt and cow dung, and she walks like she just recovered from a polio seizure."

"Okay, Mom, please stop. First of all, Jeeto does not have spindly legs or walk like she has just recovered from a polio seizure. She has beautiful legs, and her walk is deliberate and graceful like those Bollywood actresses in the movies. In fact, she walks like a fashion model even under all that weight on her head. Yes, she is always covered with grit and sh!t." I hesitated, reversed a bit, and said, "cow dung." Then I continued, "Mom, please look at it rationally. Jeeto is like an uncut diamond. She just needs to be polished. Just imagine if someone gave her a thorough scrubbing with an industrial-strength detergent and then a coating of perfumed oil, she would shine like a jewel and you would be proud to imagine her as your daughter-in-law."

"And I guess you would be the one to give her that industrial-strength scrub?" She looked at me and could not hide her smirk despite her best effort to look serious and concerned.

Mom was not about to give up. She again tried to convince me to wait and resist all such distractions. I would be rewarded later.

I had heard such promises before. Our village priest had been drilling into my head that if I was a good boy and refrained from bad things, sex being on top of that list, good things would happen to me in the afterlife. I could never take that advice seriously. Maybe I could hold off for the time frame my mother had proscribed for me, but not the lifelong commitment the preacher wanted. No way. I knew I did not have that kind of patience and, actually, neither did the preacher. A year later I found out that the preacher was accused of child molestation and sanctioned by authorities.

I did not want to ignore my mother's wishes. She had a lot invested in me. She wanted to be proud of me and my two brothers. She wanted to show the villagers that she was a successful mother.

Life had never been easy for my mother. She had been in Jeeto's place herself when she was growing up in a village not too far from ours. Even as an adult and after she married, life was harsh for her. Although my father worshipped her, he never gave her the time she deserved. He was a philosopher immersed in his countless books on religion and philosophy while my mother endured taunts, insults, and alienation from my father's side of the extended family.

My mother came from a family that was better educated than the people on my father's side of the family. My mother's parents belonged to a religious sect that many in my father's village ridiculed. She never fit into my father's family. She was much taller than the other women in the village. This she inherited from her father, who was reputed to be the tallest man in a thirty-mile radius. He became famous for his physical stature but was also the butt of jokes. On one occasion I heard a villager telling people that my maternal grandfather's knees were like the hubcaps of a British Leyland truck (popular in India at that time). Because of his height, he earned the nickname "Camel." Some people would refer to my grandfather that way even in front of my mother, who loved her father and utterly resented this insult. But somehow she restrained herself from hitting back. It bothered me very much. I wanted to punch the barbarians, but I was helpless. I was just a kid. I would just cling to my mother and cry. I wanted to help her.

My mother needed no help. She could take care of herself. She was a lady of steel—stainless steel, actually. She was more daring and better educated than her tormentors. Ultimately she had the last laugh. She started gaining respect because of her daring, intellect, and perseverance. She was way ahead of her time. She waged a vigorous campaign to bring a girls primary school to the area and succeeded. She became a respected teacher, and even some of her detractors eventually came to our house to pay their respects.

The last glance

I did not want to disappoint my mother. She did not have to worry. My summer vacation was almost over. The next day I was scheduled to leave Gurne to go back to the city to resume my studies. Somehow the whole village knew it. Rumor spread that I was going somewhere far away like America, perhaps never to return. The rumor was false. I was only going to Calcutta to explore educational opportunities.

My last day to see Jeeto was at hand. She came at the usual time. Normally she would turn her head only slightly to look at me for a few fleeting seconds, leaving me craving for more. This was different. Today she tilted her head a full 90 degrees, a painful gesture no doubt under the enormous weight on her head. And instead of a few fleeting seconds, she stared at me for what seemed like forever. Luckily (or perhaps unluckily) there were no passersby to interrupt the painful gaze full of sadness, despair, and anger. Gone was the heavenly smile, the mischievous teasing eyes, and her gentle grace. No words were exchanged, but the text was there for me to read in black ink and large font. None of it was pleasant. Jeeto's big, beautiful, black eyes were telling me in no uncertain terms: *Why did you do this? Why did you play with my emotions? Why did you lead me on if it was to end this way? You may have a lot of things in your brain, but your heart is an empty shell with nothing in it. Absolutely nothing! Go back to your world where this is just a game and it is all about winning.*

Jeeto still stared at me. I always craved more attention from her, but today I could take it no more. I wanted it to end. I wished she did not look at me. I felt bad. I felt like the green goo Jeeto was carrying on her head. Finally she turned her head with one swift jerk and continued her march to the cow dung yard. She didn't look back.

Obviously she had heard from the neighbors that I was going away the next day, possibly never to return. What was the point of smiling if there was no continuity in this play? This was just a dream, a dream with no theme. Who said life was fair?

I never saw Jeeto again. A year later when I came back to my village, Deepa told me that Jeeto had grown into a woman. Ruffians on the street had taunted her. One of them had even tried to molest her. Her parents decided she should stay home.

Deepa continued, "Jeeto no longer comes out this way. I have not seen her in months, and if you are thinking of seeing her, forget it. It will do you no good. You will only cause grief to your mother and to us all."

I heeded Deepa's advice and did not attempt to see Jeeto or even walk by her house.

Another year went by. I came home again for a couple of weeks. Still I could not resist asking Deepa if she would help me to find Jeeto only to see what she looked like then. I suggested that Deepa and I go to Jeeto's house and knock on the door, pretending to be collecting donations for an orphanage.

"Forget it, Jas. Give up. I cannot be party to any such stunt. Moreover, you will not see Jeeto anyway even if you show up at her house dressed up as a miserable beggar." Deepa continued, "Let me tell you something you may not want to hear. Jeeto is married now. Her parents found a suitable match for her in a neighboring village. She had a big wedding. I saw her husband. The man is a lot older than Jeeto, maybe fiftyish and not much to look at, but apparently he has money. They say he has 150 acres of land and a hefty bank account that he amassed when he worked in Kenya. I don't know what he did, but apparently he brought a lot of money back with him when he returned to the village."

"Okay, I know it is over. I have known it for two years. I just wanted to see what she looked like. I wanted to tell her I was sorry, but I know it is just a dream like everything else."

"I wish I had better news for you, brother," Deepa tapped my shoulder.

I doubt Jeeto is happy, but I know she will no longer have to haul that cow dung, I consoled myself.

A lifetime later

In 2010 I visited my birth village accompanied by my wife Mary and daughter Monica. At my request, my relatives set up woven jute cots on the porch where I used to sit and watch Jeeto walk by. So it seemed like absolutely nothing had changed in the village until my niece brought me her brand new laptop computer equipped with Wi-Fi. She said I should test it. Without thinking too deeply, I opened up Google, typed "Jeeto – the Uncut Diamond" and pressed enter. Hundreds of records and documents showed up, many related to illegal and immoral diamond trade in Sierra Leone. It also brought up several Jeetos, one of whom seemed to be a well-educated professional woman in India.

The Jeeto I was looking for was nowhere in cyberspace. Even Google could not find her. The diamond was to remain unpolished and undiscovered.

Monica came from behind and tapped me on my shoulder, saying: "Snap out of it, Dad! It's time to leave. The driver is getting anxious. We need to get to Delhi on time to catch our flight to Los Angeles."

CHAPTER 2

GREAT SURGEON

I met Sukhbir when we were both college freshmen in India. Like most students of that age, I was undecided about my future and even if I had a specific career in mind, could I make it happen? I wondered if I could get admission to the university of my choice. If I was admitted, could my family even afford to send me to college?

Sukhbir and I faced different decisions and different statistical odds of realizing our respective goals. I had better odds of getting into an institution of my liking because of my good grades and high test scores. However, my odds of affording an advanced education were not encouraging because of my family's limited financial resources that contained no stocks, bonds, mutual funds, real estate investment trusts, or vast acres of farmland.

On the other hand, Sukhbir had a different set of obstacles toward achieving his dream. He wanted to go to medical school and become a surgeon. He had no interest in any other career, not even in being a pediatrician. However, his chances of getting admission to a reputable medical school were not good because he was a "B minus" student. Competition for admission into medical schools was stiff. Perhaps he could have gained admission to a medical college that had lower admission standards, such as an institution in Grenada, an island in the Caribbean. His family could afford to send him to a foreign country, but he vowed to complete his education in India. He was determined to prove that he could realize his dreams in India.

Throughout our undergraduate studies, everyone called Sukhbir the "Great Surgeon" because of his obsession with becoming a surgeon. The humor of his nickname was not lost on him, and he did not object. In fact, his nickname, which was given to him in jest, seemed to harden his resolve to prove everyone wrong.

Great Surgeon came from a wealthy family in Punjab. The family had hundreds of acres of land, many farm animals, and lots of hired hands to run the sprawling business.

His oldest brother was a commissioned military officer in the Indian Army and as such made a decent living and enjoyed a high social status. He was married to a very beautiful woman who had been educated at the elite Delhi College for Women.

His other brother, who was only a few years older than Sukhbir, was a minor functionary in the regional irrigation department but a major contributor to the family's fortune. Being in the right department, he had a golden opportunity to supplement his modest government pay with monetary gifts from grateful farmers who were chronically short on rationed irrigation water for their crops.

Sukhbir's wealth was impressive, but that was not the main reason I became one of his best friends. He had other more endearing qualities. Most notable was his sense of humor. He was a very funny man. He was direct, aggressive, and sometimes crude, but he always tried to be funny.

His command of the local provincial dialect and imagination were also something to be admired. He could talk for hours about real or concocted stories. His puns and insults of people not present were the funniest. In such situations he would ruthlessly needle people without feeling the least bit guilty. Recognizing that they were not present, he did not see how his statements could hurt anyone. I had so much to learn from him, although very little of his sort of wisdom would help me in my academic or professional pursuits.

I also admired Great Surgeon for his physical stature and his general authoritative disposition. Being several inches taller than most people around him, he stood out like a sore thumb among Indians of average height. His exquisite red *Patiala Shahi* turban added another three inches to his already impressive stature. Being conscious of this, he intentionally acquired an erect posture that made him look very important. Sometimes people would salute him for no reason at all except they felt he deserved respect.

Great Surgeon's turban was a marvel of architecture. It was perfectly shaped. Not a single fold or crease was ever out of order. To make sure that the artistically designed marvel stayed intact for the intended duration, he would carefully insert three long metal pins in strategic spots. As I recall, the central pin was of fourteen-karat gold, which added luster.

The gold standard at the time for headgear was the turban worn by the Maharaja of Patiala. Patiala was a large and influential princely state in those days. Its capital, Patiala, is now an important city in the state of Punjab in Northern India.

The Maharaja of Patiala was an imposing and highly respected figure throughout India and, in fact, elsewhere in the British Empire. He had British and Indian titles that could fill half a page and were written in three separate languages—English, Persian, and Hindi.

My birth village of Gurne was inside the Maharaja's territory and about twenty miles from the capital, Patiala City. All elementary school students within the Maharaja's territory were required to memorize and correctly pronounce all the Maharaja's titles before advancing to the next level. I still remember many of those titles:

KCSI (Knight Commander Star of India)
GCSI (Grand Commander Star of India)
MEOSI (Most Exalted Order of the Star of India)

Other titles sounded more Persian. For example:

Ferzande Khaas (Favorite Son)
Raja-i-Rajgaan (Prince among Princes)
Dualt-i-Englishia (Treasure of the British Empire)

The style of the turban that the Great Maharaja wore became known as the "Patiala Shahi" (meaning "Patiala Royal"). Every young Sikh boy strove to achieve the Patiala Shahi standard. For a young college-bound Sikh boy, his turban defined his personality, his standing among peers, and his chances, if any, with the Sikh girls, some of whom were stunningly beautiful or from well-to-do families. A properly starched, good quality, precisely tied Patiala Shahi turban gave immediate status. It told onlookers that the wearer was affluent, belonged to a good family, and had a bright future. One who had mastered this turban architecture was status conscious. Truth be told, this attribute of the Great Surgeon's personality contributed greatly to his success in life.

My own turban, more middle of the road, was not too shabby. Compared to the Great Surgeon's turban, however, my turban was a disgrace. It desperately needed work. I took repeated lessons from my friend. Over time, my turban improved, and along with that my chances of meeting people and receiving appreciative glances from the coeds on campus.

To my delight, our university, part of the newly built modern city of Chandigarh in the Himalayan foothills, had a disproportionately high number of female students.

Chandigarh was the first planned city in India in modern times and was known all over the world for its architecture and urban design. The city was designed by the famous French architect, Le Corbusier. Other developments in the city were designed by the equally celebrated architects Pierre Jeanneret, Jane Drew, and Maxwell Fry. At one time the city was reputed to be one of the cleanest cities in India with the highest per capita income in the country.

The Chandigarh and Punjab University campuses were worlds apart from any other city or college campuses in India. The dormitory in which Great Surgeon and I lived would rival any modern dorm in California. We had small single rooms next to each other on the third floor. Each room had a small balcony but no air conditioning. On hot nights we would sit on chairs on our respective balconies and tell each other tall tales, giving no regard to the decibel levels of our voices and oblivious to how we may have disturbed our more studious neighbors.

Every Indian family that had the means to put a child through a graduate-level education wanted their child to become a medical doctor. The professions of scientist or engineer were lower in status. The engineers and scientists were respected for their knowledge and intellect, but only doctors made big money. Doctors made a good living even if they chose to practice their trade in the rural areas where the villagers did not earn much money. The sheer number of patients who needed basic medical attention was so high that a doctor willing to work long hours could fill his or her coffers even with the modest fees they charged.

Becoming a medical doctor was a big thing in my family, too. My mother wished for me to be a doctor. My father never stated his preferences for my career. He just wanted me to do my best. He was generally aloof from such materialistic pursuits. He delighted in reading poetry and understanding the nuances of ancient Sanskrit writings so he could savor the utterances of the sages in their purest form rather than through a half-literate translator. He also studied the world's other great religions. I was amazed at his understanding of Christianity and Islam even though he had no personal exposure to these religions while growing up. He even learned some rudimentary English words to help him better understand Western civilizations. I know that my father would have been equally happy if I had become a poet or a musician. He believed I had musical talent and an ear for music, which I do.

I never became a medical doctor, which I thought must have disappointed my mother though she never expressed any displeasure. She was too sweet. Her heart swelled with pride when she found out that I was going to America on a US Government scholarship. She figured I must be important because the Americans agreed to fund my graduate education in the United States through a US Air Force research scholarship. I know she was proud of me because the day I showed her the letter from the University of Southern California (USC) offering me the fully funded government fellowship, she immediately grabbed the crisp letter from my hands. She went to each and every house in the village, knocked on the door, and told the neighbors, "My son, Jas, is going to America. Americans really want him over there so much that they will finance his studies while in the United States, give him a nice place to live, and good food to eat."

As proof of this, she would thrust the letter from Professor Anton Burg into their face knowing full well that, just like her, none of them could read a word of English. That really did not matter.

My mother did not live long enough to see my success despite my not having chosen a medical career. Before she died, she did realize, however, that people called me doctor. Understanding how anyone can be called a doctor without having studied medicine was difficult for her to grasp. I wish she was alive today.

The chemical doctor

The "Chemical Doctor" label has landed me into some interesting and very funny situations. I don't go around promoting the "Dr." title in front of my name, but I have found that when travelling, using Dr. does give you some advantages. Airlines like the idea of having a doctor on board even though they don't always know what kind of a doctor is there. Unfortunately, on long flights with hundreds of passengers, someone always gets sick. Having a doctor in the cabin can be handy.

On one such occasion I was approached by a flight attendant, saying, "Dr. Singh, we have a lady up front who is not feeling well. As you know, we still have three hours of flight left. Would you mind taking a look at her and advising accordingly? In fact, if you wish, we can assign you a seat up front in the first class section so you can sit close to the indisposed individual for the rest of the flight."

The idea of moving up front to the first class cabin was very appealing, but not to the point of masquerading as a physician. I had to tell the flight attendant, "I am sorry, but I am unable to help because I am not a medical doctor."

"So you must be an academic, a PhD?" the flight attendant inquired.

"Yes, I am that kind of a doctor—a doctor of chemistry, actually. I wish I could be of help to the lady in distress, but I cannot."

"Thank you, sir. Sorry to disturb you," the attendant said as she walked away.

A doctor for the Quebec beauty queen

It was late in June of the second year my wife and I were living in Ottawa, Canada. I had just finished my two-year research fellowship with the Canadian National Research Council (NRC), a marvelous research institution in Ottawa. The NRC was such a prestigious institution that recruiters from major corporations came in droves to the NRC campus to interview prospective employees. After a couple of interviews, I accepted a job offer to be a researcher in the lubricants and additives research laboratory at the Amoco Oil Company in Chicago. I was very excited. This was my first real job paying real money after being a post-doctoral student for four years. I do not want to imply that I did not value my academic research fellowships. To the contrary, during those years I had the best time of my life and learned a lot. My wife and I were very excited at the thought of going to Chicago, buying our first house, and raising a family. Our yearly lease for the small one-bedroom apartment in East Ottawa was ending.

We packed to go to the United States. The problem was that I did not have a visa to live or work in the United States. I had applied for it in a timely fashion, but it was not forthcoming. I was told such visas could take three months or more.

The two of us decided to stay in a small motel outside the Ottawa city center until the coveted piece of paper was ready. We had a small room with an efficiency kitchen. Next door lived a young French Canadian couple. The husband, Marcel, was a taxi driver. He worked very long hours. I rarely saw him as he left early and came home late. My wife and I spent our days and nights in the motel. We had nothing else to do but wait for my visa. And wait we did. During this time we became friends with the taxi driver's wife, Rose. She also had nothing else to do when her husband was plying his trade day and night.

Rose told us that in her youth she was considered the prettiest girl in the area and had won multiple beauty contests in her hometown in rural Quebec. We did not doubt her words. She could still stand out in a group of women even fifteen years after her beauty contest. Rose and my wife became friends. They swapped recipes. They planned "cook-ins" in our respective motel rooms. Rose prepared for us her all-time favorite dish "Coq au Vin" (chicken breast cooked in wine). It was exquisite, every bit as good as you would get in a Michelin three-star restaurant.

Because of Rose's friendship and cooking, the long days at the cheap roadside motel were not boring or lonely. Our fear of developing cabin fever vanished.

One day Rose knocked at our door. She was excited. She had gone to the motel manager's office to pick up the mail, a daily routine. While there, she also picked up our mail. The address on one of the letters read, "Doctor Jas Singh." Rose handed the envelope to my wife and said, "Mary, you never told me Jas was a doctor."

"Rose, I did not think it was important. We are friends. It does not matter who is what. He usually does not broadcast the doctor thing unless it is required at his work. Sometimes he does mention it when he is booking his travel. He thinks it gets him better treatment, maybe a better seat."

"I am sure it works," Rose replied. "You must be very proud of him. He is such a gentleman!"

At this, Rose moved closer to my wife. She lowered her voice, which was still clearly audible to me, and said "You know, Mary, I have had these health problems and I think I might be pregnant. Marcel works a lot, but we still do not have a lot of money. Do you think Jas would mind giving me a checkup?"

Mary had a faint smile on her face. With a straight face she said, "Rose, I don't think Jas would mind, but I would."

Rose was taken aback. "Why is that, Mary? Is he not a doctor? Is this not part of his job? You don't trust him, Mary?" Rose was clearly troubled by my wife's statement.

"No, Rose. It is nothing like that. Of course I trust him. He just is not that kind of a doctor."

"Then what kind of a doctor is he?" Rose asked. She was still troubled. She felt Mary was implying something negative related to her character.

"Rose, the best way to explain this is that Jas is a chemical doctor, the kind that teaches or does research at a university. He is a PhD."

"I guess I have heard of that kind of a doctor too, and I am sure you are still proud of him," Rose reassured my wife. Both women then burst into laughter at what had just transpired.

Western medicine or faith healing

Even though I pursued a different track for my career, I was surrounded by medical people. Both of my cousins—who grew up with me after their mother died during childbirth—went to a medical college. We were very close in age. Both were girls, and because they lived with my family early in their youth, I always thought of them as my sisters. The older of the two girls graduated with what was called the MBBS degree from the Christian Medical College in Ludhiana, Punjab. My younger cousin graduated from the same institution with a public health nursing degree.

The Christian Medical College and Hospital in Ludhiana, India, was the first medical school for women in all of Greater Asia. It was founded by Edith Mary Brown and has since been renamed the Brown Memorial Medical College. The foundations for this venerable institution were laid in 1881 by medical missionaries, the Greenfield sisters Martha Rose and Kay. They were evangelists and educators from Scotland. The Brown Memorial Medical College was officially recognized by the Indian government in 1915.

My cousins benefited enormously from attending a medical college that exposed them to Western cultures and rigorous educational standards. Their horizons were expanded and they both had fulfilling careers. After graduation, my oldest cousin married a man who had graduated the same year from another medical college in the area. Rumors were whispered that theirs was to be a "love marriage." At that time in India, most marriages were arranged by the family regardless of the bride's education or interests.

My older cousin and her husband's love marriage was a match made in heaven. Throughout their life together they were very happy. Both of them were funny and smart. My cousin was smarter than her husband. Her husband, a tall and handsome man with inquisitive eyes (which made some people nervous at times), was much funnier than my cousin.

As one who enjoyed his good sense of humor, I loved to go to his clinic during my summer vacations to watch him ply his trade. Patients' privacy, as known to us, was not an issue in that place and time. A doctor could discuss any detailed diagnostic matters with a patient while other people listened. I found such medical details educational and entertaining. My listening to such personal patient-doctor discussions was not a problem. I was seen only as a kid, and kids never mattered when adults were talking about important things. The only time I remember being asked to leave the room and have tea with a young female nurse, which I appreciated, was when the doctor discussed a sexual dysfunction with a male villager. As a matter of policy, he never discussed such things with female patients, as that would be the task of his wife, my cousin, who was a consultant to his clinic in addition to her own work in another clinic.

Patient restraint

One afternoon I was sitting in my brother-in-law's office telling him about my studies at school. It was a busy day. A man came in, bowed to the doctor, and sat down on the bench. The villager looked at me and said to the doctor, "Nice looking chap, Doctor Sahib. How old is your son?"

"He is not my son. He is the lady doctor's young cousin," he told the villager. The lady doctor he referred to was his wife, my cousin. He always called her the lady doctor except when the two were alone.

The villager told the doctor about his distress. He seemed to have a cold accompanied by a dry cough and runny nose. He felt he had a fever but was not sure. He talked nonstop even through bouts of coughing.

"Open your big mouth as wide as you can so I can place the thermometer properly," the doctor ordered.

The villager complied. The doctor inserted the thermometer and gave him instructions not to take it out until told to do so, even though it would be uncomfortable. While waiting, the villager took off his brown wraparound blanket, a *khase*, and settled down to receive his next command from the doctor. Two minutes passed. The doctor motioned to the patient to continue to wait. Another patient walked in and received attention. The man with the thermometer stuck in his mouth grew uncomfortable. He was also cold now without his khase. Ten minutes elapsed. The poor man's discomfort increased. The doctor motioned for the patient to wait for five more minutes by showing his five fingers of his left hand while his right hand was busy on another patient. By now the man was desperate. He began making funny gurgling noises. His eyes were bulging. He attempted to draw the doctor's attention by sounding like a seal. He wanted some attention from the doctor but would not dare take the thermometer out of his mouth.

Finally pulling the thermometer from his mouth, the doctor ended

his agony and patted him on the back. He said, "Good news, Tiger! You are as healthy as a water buffalo bull. You just have a little nasal congestion. Don't go into the fields today. Go home right away, get some rest, and take this white pill with some milk. Take another one just before going to bed."

"*Shukrya* (thank you), Doctor Sahib, for your patience." The villager looked at me again and said to the doctor, "Please tell the lady doctor Sahib that her young cousin is a fine boy."

The man left, but before he cleared the clinic gate he came back and said to the doctor, who had just turned his attention to the next patient, "Sahib, as for the white pill that I am supposed to take before going to bed, it is unlikely that there would be milk at that hour. Any milk would have turned sour by then in this godforsaken hot weather."

"You can take the pill with tea and no harm will come," the doctor assured him without ever looking at the man. The villager thanked him again, this time saying shukrya twice.

Apparently he still had some doubts. He came back a third time and said, "Sahib, I was thinking, there is no certainty that there will be any tea at ten p.m. when I am ready to take my night pill. Will it be acceptable if I swallowed the pill with water?"

At this time the doctor was in the middle of examining another patient. He had the patient's eyelids parted wide open to look at his cataracts. The doctor lost his cool at these persistent interruptions. He shot back, "Look, mister, the effect, if any, will be from the chemical in the little pill. As for the accompaniment, I don't care if you take it with piss! Now please go home and let me do my work."

"Sorry, Sahib, sorry," the man said and exited in a hurry. He was still murmuring, "Sorry" as he cleared the gate.

At lunch time when there finally was a breather, I asked, "I have two questions, if you don't mind answering."

"Okay *Beta* (son), shoot. What do you want to know?"

"We both know that fifteen minutes are not required to take someone's temperature. So why did you gag the poor man by asking him to keep the thermometer in his mouth all that time?"

With a laugh the doctor asked, "Okay, what is your second question? I will respond to both questions together."

"If how you take the pill has no therapeutic significance, then why would you prescribe the pill with milk in the first place?"

"Excellent questions! I will tell the lady doctor that you are destined for something big. You are a very perceptive young man. You belong in America, not in this hell hole."

"Now for your first question. Why did I keep the thermometer in the man's mouth so long? That is the only way to keep the man from talking constantly. I know him well. He comes here because he gets bored or lonely at home. The guy just talks and talks and never shuts up. This is disruptive and is not fair to the other patients. I guess I could use my medical prerogative and strap his jaw shut with an adhesive tape under the pretext that this is part of his treatment, but that could be construed as medical malpractice."

I was then going to ask him if putting a thermometer into someone's mouth for fifteen minutes could also fall into the malpractice category, but I kept quiet. I did not want to jeopardize my "Beta" status.

"As for emphasizing the need for milk to accompany the medicine, I find that milk has psychological value. I don't know if you have heard of the 'placebo effect.' Don't forget, milk is revered in North India. It is the elixir of mother earth, nature's gift to man, and if you are a religious man, milk adds another curative factor. The nectar provided by the 'Go Mata' (mother cow) can have a useful medicinal effect. Some patients think they are getting two treatments in one. Some people actually get well even if the pharmaceutical I give them has little or no biological potency. There was nothing wrong with the man who got me all riled up this morning when I was so busy. I could have given him a capsule full of sugar and he would tell me the next day that it had magical effects and would thank me profusely."

I was impressed with the explanation. I figured that you had to be a psychiatrist and a faith healer in addition to having Western medical training to be a successful family physician in rural India. I spent three more mornings in the clinic. Each day with my cousin's husband was more enlightening than the last. Finally I had to say goodbye. The time had come for me to return to school.

Great Surgeon rediscovered

Fifteen years had passed since I had left India. My children were finally old enough to endure travel in India. My son David was nine years old and my daughter Monica six – both old enough to enjoy the sights and sounds of India, bond with their cousins, uncles, and aunts, and learn a few words of the language of their father. I was afraid that the local vocabulary they would pick up in my village would most likely be the words they should avoid, but I did not worry too much. I knew that after a couple of months in Michigan, they would lose these literary acquisitions because they would have no way to practice those words. I was not prepared to help them retain their knowledge of curse words.

Our family visit to India was going well and exceeded our expectations. One evening I decided to take some time away from my family and go visit Great Surgeon.

Great Surgeon had beaten all the odds. He studied day and night and improved his test scores to the point that he was accepted into a medical college in Punjab. He graduated from medical school and found a government job as a physician in a special family planning unit where he quickly moved up the ranks. When I met him he was performing his surgical wonders in a town about one hour's drive from my boyhood home.

Great Surgeon was the head of a medical unit assigned to implement family planning programs in the area. Birth control was a major priority at that time for Mrs. Indira Gandhi's government. As I recall, the process was that a team of medical professionals—including a doctor, an assistant who usually was a paramedic, and a social worker—would visit every village in their assigned territory once a year. Their initial visit would involve gathering the villagers, explaining to them the benefits of smaller families, describing various forms of contraception, and giving the villagers a generous supply of free condoms or the option of a vasectomy if the man already had a couple of children. The men who opted for a vasectomy would have the procedure completed at a proper medical facility at no cost. While a vasectomy was the gold standard of success for contraception, further incentives were offered to encourage the men to submit to the surgical procedure.

A man who agreed to a vasectomy would receive a brand new AM-FM radio. The imported electronic device had strong enough reception to allow him to tune into the "Binaca Geetmala" (necklace of songs) program from Radio Ceylon. Radio Ceylon was the only private, non-governmental, broadcast facility within the listening area and would broadcast pure entertainment interrupted by some commercials but without the annoying government propaganda and constant ethical and moral sermons that nobody paid any attention to anyway. The AM-FM radio incentive was very powerful, more so than speeches and posters.

Great Surgeon had become a specialist and a leader in the crusade for male sterilizations. While most of his colleagues in the family planning unit were primarily involved in distributing free condoms, Great Surgeon performed surgeries—his long dreamed-of specialty. He performed hundreds if not thousands of vasectomies. He may have performed other surgeries too, but I don't know about them.

Great Surgeon was itching to tell me about his exploits and some very funny stories. For a warm up, he suddenly yelled at the boy who was one of his domestic helpers, saying, "Mr. Idiot, bring some ice-cold beers for the American Sahib and me."

"Yes, Doctor Sahib," the boy replied in perfect English. "Right away, sir!"

I was surprised at the harsh words and said to the Great Surgeon: "Why did you call the kid an idiot? He is probably smarter than 90 percent of the kids around here. He is a handsome kid and he is polished. Doesn't he resent being called idiot?" I asked.

"On the contrary, he likes it," Great Surgeon explained. "He likes it better than the name he was called by his previous employers. They called him *mundoo* and Mr. Idiot hated it. First of all, mundoo was not an English name, which is preferred over the *desi* (domestic) name. Second, not even the locals know what mundoo means. The word *munda* means a boy, but mundoo, the diminutive form, does not make much sense. The best I could figure out was that mundoo is a boy in the formative stage that has not yet attainted any status. So you can see why he likes his English name, 'Idiot.' His friends are jealous of his English name. To tell you the truth, he is like family. He eats with us, a treat not accorded to household help in India."

The discussion about the nomenclature completed, Great Surgeon went on to tell some of his best vasectomy stories.

A small family is a happy family

Great Surgeon, his paramedic, and the social worker went to a village in the southern part of the Punjab State. The team gathered the villagers in the middle of the town. The social worker explained in great detail about the benefits of having a small family. He told them that villagers could afford to educate one or two kids but not five or six. With smaller families they could provide better life opportunities for their children. Colorful posters showed the villagers how families with only two kids looked happy.

After the social worker, Great Surgeon had his turn. He told villagers of several ways to limit the size of their families. Prophylactics such as condoms drew a lot of interest. Great Surgeon saved the best for last. He told the gathered young men that if they already had two or three kids, they should have no more. They should consider a vasectomy. It was a guaranteed procedure. It required only a minor operation, and in the hands of an expert like him, it did not hurt at all.

The idea of a surgical procedure frightened many in the audience. Their main concern was that perhaps the procedure could adversely impact their manhood permanently. Being a master communicator, Great Surgeon erased all fears.

At least ten young men opted for the procedure, and eight of them later went under Great Surgeon's knife at his clinic. This was a tremendous success. Records at the clinic bore out the successful statistics in terms of fewer pregnancies in the area.

A year later the same team went back to the same village to measure the progress. Great Surgeon went through his routine. After an impressive speech, he asked for a show of hands as to how many thought the program was successful. Quite a few hands went up. The program had been successful.

Tea and sweets were served, and the team completed the necessary paperwork and prepared to leave.

As he was leaving, an old man caught up to Great Surgeon, pulled him aside, and said in an angry voice, "Your speech was impressive, but your surgical skills leave much to be desired."

Great Surgeon was stunned. "What do you mean my surgical skills leave much to be desired?" he challenged the old man. "Didn't you see how all the villagers gave thumbs-up to the program?"

"Maybe, but what I want to tell you is that you performed a vasectomy on my son last year, and now his wife is expecting a baby any day. So don't tell me about your surgical skills and your gold standard." The old man was getting outright hostile and abusive. Others could hear him.

The outburst exceeded the Great Surgeon's limit of tolerance. He had performed the surgery himself. There was no way he could have botched the operation. He was not at fault. The old man was out of line. Great Surgeon shot back, "Look, Grandpa, before you go accusing others, take a look into your own household. I operated on your son. I did not sterilize the whole village."

The old man was speechless. He shook his head and left.

Great Surgeon then burst into a thunderous laughter, realizing the potency of his own joke and the success of his mission. He asked Mr. Idiot to bring more beers.

By then Great Surgeon was on a roll. He went on to tell another family planning story. The mission in this case was understandably not very successful. According to Sukhbir, this work had not been carried out by the Great Surgeon's team.

"Tell me and I will forget
Show me and I will remember
Let me do it and I will understand."
　　　　　　　　　-Kung Fu Tze

A team of three medical professionals went to a remote Indian village. The place was backward even by the standards of the region and the time. The team gathered the villagers and went through their rehearsed routine. The male social worker outlined the issues, gave a well-rehearsed pitch on the virtues of raising a small family, and even gave specific examples of how each extra child would crimp the family budget and strain the finances of an otherwise financially sound household.

Next it was the doctor's turn to explain the available means of family planning. His speech went smoothly. The paramedic prepared to distribute the latex devices the doctor had recommended for the men. "Before we give these supplies to you, does everyone understand how to use the devices?" the visiting doctor asked.

There was total silence. The doctor repeated the question. Still no hands were raised. Finally one brave individual spoke up: "Doctor Sahib, you explained well, but many of us here are still not sure how to use these gadgets. Perhaps one of you could demonstrate it for our benefit."

The doctor, the paramedic, and the social worker looked at each other. They seemed helpless. There was no way to do this. Suddenly the social worker had an idea. "Someone please cut a branch from a nearby mulberry tree about one-and-one-half inches in diameter," he said.

Five or six villagers dispersed in different directions to fulfill the requisition. Within minutes, they had fetched half a dozen mulberry sticks with the prescribed dimensions. The paramedic stepped forward, carefully slipped the condom on the stick, and declared, "This is all there is to it, folks. Can everyone do this?" Everyone nodded vigorous approval.

One villager reassured the medical team, "Gentlemen, we are simple folks. We don't have college degrees and we cannot read all those English labels, but what you showed us is so simple that any idiot can do it." The medical team applauded, promised to return the next year, and left.

The next year on the appointed day, the same team returned, gathered the villagers, and posed the all-important question, "Folks, we were here last year. We had good discussions and a good understanding. Before we leave you with fresh supplies again, we must determine if the process is working as intended? Are we getting the desired results?"

Dead silence followed. No one raised a hand. Something was amiss. The doctor repeated, "Is the process working? Are there fewer pregnancies in the village since last year?"

One hand went up. It was a senior elder. He shook his head in a negative manner and described the problem this way, "Doctor Sahib, the process is working fine, but it is not effective. The devices are not complicated. Even the uneducated can do this as well as the smart ones. The problem is that the women in the village are still becoming pregnant with the same frequency as before even though we covered scores of fresh mulberry sticks with the latex you provided. All were made to your specifications."

The medical team members looked at each other in dismay. The doctor was heard saying to the other two, "Fellows, we need to regroup and rethink. We should focus our strategy more on the vasectomy campaign or improve our communication skills."

In reflection, Great Surgeon let out another roar of laughter, this time of longer duration. The failure in this case was of another team's, not his. The story he told me was most likely true, as it had been reported in the regional press. It was possible that Great Surgeon was the one who led the team that failed to communicate the proper way to don the condoms, but he would never admit that this was his failure. If he did, he wouldn't be the Great Surgeon!

CHAPTER 3

SILK PAJAMAS

Shanghai is one of my favorite cities. The massive city of twenty-three million people is crowded, chaotic, and vibrant. There is never a dull moment. People from Shanghai are inquisitive and nosy. Yet you can find instant acceptance if you can cross the language hurdle, which is much easier said than done. What encourages me about Shanghai is that if you can master just a few sentences, and especially if you can learn to write a sentence or two (legibly that is) in Mandarin, you will take a quantum leap forward in your quest to connect and make friends. Like most other people, the Chinese love their native tongue.

A Chinese friend who lives in Singapore agreed to teach me the two Mandarin sentences I coveted the most. I was tired of just hello 你好 and thank you 谢谢. After weeks of practice I could write the two sentences my Singaporean friend had taught me:

你很美

(You are very beautiful.)

Once I mastered the beautiful part, I graduated to:

我喜欢你

(I like you.)

The very first time I wrote the "You are very beautiful" sentence was to a receptionist at the Sheraton Hotel in Wuxi. This produced great results, which included an upgrade from the cheapest room in the house to a suite.

Buoyed by the experiment, I used this sentence shamelessly to my advantage at every opportunity, once writing the sentence (albeit absentmindedly) even to a man. This produced some unintended and unwanted results.

Souvenir hunt

The first time I went to the silk capital on a business trip, I was determined to buy a present for my wife. I am a lousy shopper when it comes to choosing gifts, but I figured it would be easy. I reasoned: what is the first word that comes to mind when you want to buy a gift for your wife in Shanghai? Silk, of course. Silk conjures up images of luxury, beauty, and romance.

Legend has it that silk was first developed in ancient China in 3500 BC by a Chinese empress, Leizu (Hsi-Ling-Shih, Lei-Tzu). Reserved originally for use by the kings of China for their own use, the fabric rapidly became popular because of its texture and luster and became a staple of pre-industrial international trade. The emperors of China strove to keep the precious sericulture secret to maintain their monopoly.

I decided that I would buy something made from silk for my wife, maybe silk pajamas, which would be a perfect gift to show her that I still appreciated her. The hotel concierge gave me names of several places but finally zeroed in on one. She said the staff at this store spoke English and were very helpful in selecting the right item if you were a man uncomfortable with picking out women's clothing.

The trip to the silk center went smoothly. The sales lady asked me to describe my wife so she could pick the right garment. I was not used to describing my wife's physical characteristics. Instead, I showed her a photo and told her my wife's color preferences. She wanted to know my wife's age, which I could not remember at the time.

"Not a problem; your wife is the same height as I am and probably wears the same size clothes," the sales lady assured me. She then pulled out several styles of silk pajamas in different colors and patterns. She would swiftly unpack each one, hold the piece against her torso, and ask me if this was her color and also the kind of patterns she wore. I was supposed to have such information at my fingertips, having been married to my wife for three decades, but I suddenly went brain dead. I asked the sales lady to choose one, having seen the photo. She selected saffron silk pajamas, held against her abdomen, and asked:

"Should I show you the top?"

"The top? I don't think so. I am not sure if my wife likes tops, although tops look good," was my response (I was still brain dead). She did not tell me anything else about the top. "Maybe next time I come to Shanghai."

"As you wish," she said curtly in perfect English, handed me her card, and thanked me for visiting her shop.

The saffron pajamas were a hit, but my wife told me that I should have bought the top, too. She hoped that the matching piece would still be there when I went back.

An oasis spelled 'Marriott'

I do not own shares in Marriott Hotels.

At last count, the city of Shanghai had at least ten Marriott hotels, most of those belonging to its Renaissance brand. Shanghai Renaissance hotels are my favorite. Priced lower than the full Marriott's, Renaissance hotels offer similar comfort and amenities at a more reasonable price. Of special value are the rooms on the concierge floors because they come with free access to their executive lounges. Shanghai hotel lounges are not like the kind you encounter at some hotels in the USA. I recall going to a hotel lounge a couple of years ago in New Orleans. The two hostesses were deeply involved in a loud argument and never noticed me. After a long wait, I finally interrupted their discourse and asked for a beer.

"It is not time yet. Be patient, sir," the middle-aged, hostess admonished me. I felt like an intruder and exited in a hurry.

Shanghai hotel lounges are different. Staffed with attractive, attentive, and bilingual hostesses, they are like oases for foreigners gasping for a chat in a language they can comprehend. I am sure Mandarin is a beautiful tongue, but after hundreds of "nǐ hǎo's" (**你好** – hello) and Xie Xie's (**謝謝** – thank you), a foreigner craves a tongue in which he or she can converse. Marriott and Renaissance hotel lounges offer a needed respite. Lounge hostesses are eager to engage in conversation with the foreigners to improve their own English. Many are students at local universities studying subjects like international commerce, management, marketing, and advertising. They fall over each other to give out information on places to go, bars to try, and shopping destinations. On your lucky day, one of them might even offer to go shopping with you.

I have stayed at almost every Marriott property in Shanghai, including the magnificent JW Marriott Hotel, but Renaissance hotels are my favorite, especially the two that are close to our office. During my fourth stay at the Renaissance Hotel near our office, after a long day, I headed straight to the concierge lounge. As I sat down, hostess Wendy appeared sporting a big smile as always.

"The usual?"

"Yes, the usual. I've had a hectic and frustrating day," I sighed.

Wendy brought a very full glass of Chardonnay and some dim sum. She told me she was doing well with her studies and in one more year she would have her degree and was already looking at some American universities. She quickly rattled off names of major US universities and was exploring ways to finance her studies in America. She really wanted to go to the US. "It looks cool and inviting." She told me that of all the places she had seen on TV, Hawaii looked the coolest.

"Actually it is quite warm," I interrupted, trying to slip in a clumsy joke. The joke was completely lost on her, and she continued, "Maybe sometime you will tell me more about it and oh, do you have a girlfriend in Hawaii? I heard there are many beautiful girls in Hawaii."

"Yes, I have a girlfriend in Hawaii and she is very nice."

"You will impress her if you give her Chinese presents. The best one will be something silk. Shanghai is world famous for silk, you know. I know men are not very good at selecting women's clothing. Only a woman knows what another woman would like."

I told Wendy that on my last trip I gave my girlfriend (my wife) silk pajamas but only the bottoms. "I don't know why I did not give her the top. I was not thinking, as usual. Now she wants the top, too. The pajama bottoms were such a hit that I might buy one more but maybe in a different style."

"I can help you. I know the best places in Shanghai. I have a real eye for fashion and I can bargain for you." She continued, "Maybe I can model it for you—the top that is— because you said she already has the bottoms. I have this Saturday off."

"That is very generous of you, Wendy, and how could I refuse such an offer, but I must go back on Friday, and Friday is your workday anyway. Why don't we go shopping on my next trip to Shanghai? I am coming back in six weeks."

"That will work. Send me an email message before you come, and I will find time to go shopping with you."

Wendy left for a few minutes to tend to a customer who had just walked in. After couple of minutes, she returned.

"I don't think I ever gave you my email." She grabbed a piece of paper and wrote:
"—k96@QQ.com."

I promised to send an email to Wendy before my next trip to Shanghai.

On Friday, I decided not to go to the lounge. Contrary to what I had told Wendy, I was not leaving on Friday. I was leaving on Saturday. Going shopping with Wendy so she could model silk pajamas for me seemed too costly. I was not willing to pay the price.

QQ is a no no!

Later, I could not resist relating the conversation with Wendy to William, my colleague and confidant in the Shanghai office. He seemed amused and said, "Don't get too excited, Jas. Those girls can be very flirty. I bet they are that way with everybody." He tried to deflate me in a sincere effort to keep me away from trouble.

"Okay, William, maybe you are right, but I must tell you that Wendy spent far more time with me than any other guy even though some of them were younger than me."

"Well, then that explains it," William said quietly like he always did when he discovered some obvious answer to a puzzle.

It took me a while before the full significance of the Confucius-like pronouncement sank in. William was implying that those hostesses usually hang around aging men.

Still refusing to let it go, I pulled out the piece of paper Wendy had given me with her email address. William looked at it and froze. His demeanor changed. He seemed genuinely worried about me, his mentor.

"Jas, you are playing with fire. You should not have this email. Get rid of this." William was giving me commands, an unusual behavior for him.

"Playing with fire? Playing with fire?" I repeated. "Getting an email address from a college student working nights to support her education is what you call playing with fire?"

"No, not that, but did you see the address that says QQ?"

"Yes, but what is so dangerous about that?"

"QQ is an instant messaging computer program. It offers chat rooms, games, personal avatars (similar to 'Meego' in MSN), Internet storage, and Internet dating services." William felt compelled to educate me on this massive Chinese social network.

He continued, QQ had more than seven hundred million accounts for instant messaging in 2011. The program, sometimes also known as "Tencent QQ" is maintained by a Hong Kong company called Tencent Holdings and ranks eleventh in the world behind only Twitter. The QQ culture is so dominant in the lives of hundreds of millions of young Chinese people that the Q coin has become a virtual currency used by QQ users to purchase merchandise and QQ-related items.

Because of QQ's extensive use of advertisements, it has been branded as malicious adware by many anti-virus and anti-spyware experts. It was this aspect of QQ that worried William the most. A secondary concern was the possible corruption of the morals of his mentor.

I assured William it was all just a game. I had no intention of sending QQ messages to lounge hostesses in Shanghai.

Silk pajamas

Four weeks lapsed after my Shanghai trip. Unexpectedly, one day I noticed an email message that somehow miraculously survived the anti-spam guillotine. The sender had a QQ address. I instantly knew it was Wendy. No one else from the seven hundred million QQ addicts would know me. But how did Wendy get my email? Yes, I did promise to send her names and whereabouts of some US universities that offered degree programs in hotel management, Cornell being one.

Against William's sound advice, I reluctantly opened the email. It read:

"Hi Jas, It was so nice for you to talk to me. You are a very interesting man and although you are older than me you look very young. I was thinking: So nice that I have a friend in America, a nice friend. Last night I was watching a show about Hawaii on TV. It looked so beautiful. In Chinese, we say 'mei boo sheng shou.' I decided I want to see Hawaii now that I have a friend in America. It seems very far. I don't know. My girlfriend told me it is very hard to get a US visa but I think if I get admission to Cornell University you told me about, then I can get a visa. I am thinking I will first go to Hawaii for few weeks and then to Cornell. Maybe you can, how do you say, 'Americanize' me before I go to Cornell. Before leaving Shanghai I can shop for you and bring silk pajamas for your Hawaiian girlfriend. I am sorry but I must tell you I feel jealous."

I could not resist imagining returning home from my next trip with Wendy at my side.

My wife and I have a sort of a routine we go through whenever I come back after a long trip abroad.

"So you decided to come back home," she will say.

"Of course, who else will have me at this point?" I will respond, and then we will burst into laughter.

"You are always missed and welcome, and by the way did you eat anything on the plane and did you bring me something from Shanghai like that silk top you forgot the last time?"

"Yes, wait till you see. Not only did I bring silk pajamas for you, but this time also someone inside the silk pajamas as a souvenir."

At this time my fantasy abruptly ended. I did not want to fantasize any further.

Change of venue

I never replied to Wendy's QQ e-mail. William's "playing with fire" words flashed like a warning beacon in front of my eyes. Playing with fire is not my style. Years of risk management training has taught me to always stay at least two standard deviations below the "safe" threshold. (In my profession, standard deviations are permitted and even encouraged in many situations.)

William's advice was ringing in my ears. For my next trip to Shanghai I decided to stay at a different Renaissance hotel that was equally close to our office. I deleted the QQ address. Wendy never received information from me on the Cornell School of Hotel Management. I felt bad, but I could rationalize. Wendy could find the information on her own. She had the QQ.

CHAPTER 4
WHAT A PROFESSION!

I am an industrial hygienist.

Industrial hygiene is what I do, and I make a good living doing this work. I have friends who are more technically proficient than me and they make a nice living, too. The problem is that it is not easy explaining to others what we do. Try telling your next door neighbor that what you do allows you to own a fancy foreign car and to live in a nice house next to a famous retired medical doctor, a successful real estate agent, and a stock broker. I have tried to explain my profession to others on many occasions and have failed. You cannot adequately explain industrial hygiene (IH) to non-industrial hygienists. On one occasion, a friend who located my house on Google Earth remarked: "Jas, I can see your place with lots of trees on the Big Island from the address you gave me. Did you work for the Mafia to earn this? You could not have paid for this on an industrial hygienist's wages." I assured him that IH had been good to me and I had been working double shifts.

I have a hard time understanding why many of my colleagues feel that we need to call our profession something different to gain respect. Why do they feel like Rodney Dangerfield, the celebrated comedian whose famous words were: "I don't get no respect?" The search for "respect" that was never missing has consumed much of our time and energy.

The Brits call this profession occupational hygiene. Americans have tried to copy the British in a vain effort to make the term more understandable. I remember the cantankerous debates and the pitched verbal battles at the annual AIHA (American Industrial Hygiene Association) conferences to change the name of our profession. I suppose all this was in an effort to garner better understanding and respect, but it amused me. Why don't we simply say: "I am an industrial hygienist? I work to make the workplace safer and healthier and protect lives." What can be more respectable than that?

While the name battle goes on, the profession of industrial hygiene keeps flourishing, having found a home in more and more countries beyond the shores of North America and Europe. The reason is obvious. The world needs to take care of its workers who fuel the global economy. The world needs industrial hygiene (okay, occupational hygiene!).

While I do not see any need to change the name, I do agree that the name "industrial hygiene" does provide for some misunderstandings and hilarious situations.

The dental clinic in Moscow

In 1984 I joined a delegation of industrial hygienists destined for the Soviet Union as part of a professional exchange. The delegation was led by Mr. Chain Robbins (now deceased), a prominent industrial hygienist with US Steel in Pittsburgh, Pennsylvania. The purpose of the three-week trip was to visit six Soviet republics and the associated local organizations presumably engaged in IH activities. The visit included academic, industrial, and governmental institutions. The idea was that our delegation would make presentations on IH activities in the United States and our hosts would reciprocate.

After two days of sightseeing in Moscow, it was time to go to our first industrial hygiene professional visit and exchange. Our tour operator told us: "Tomorrow, after breakfast, we will take you to meet with your hygiene peers at the Moscow Medical Clinic. There you will hear about the activities of the clinic and the equipment they use. You will exchange information with the clinic staff and can ask questions. We will then have lunch. Afterward we will drive leisurely past the Kremlin and the magnificent St. Peter's Basilica and then take you back to your hotel."

That sounded good. The fact that we were visiting a medical clinic did not dampen our enthusiasm. We were aware that industrial hygiene was not an established discipline in Russia or in the rest of Eastern Europe. Industrial hygiene was buried deep under the clinical medical practice or was a subdiscipline under public health, which was also all medical. Knowing this made us even more curious as to what we were going to witness.

After a slow and long drive, we arrived at the clinic. At the entrance to the facility we were met with a warm welcome by the clinic staff that included bouquets of chrysanthemums. After the pleasantries were over, we were led to a "theater" where we screened several children's films that all showed healthy, eager, and smiling children. This was followed by short lectures on children's dental diseases by two of the senior physicians at the clinic. Apparently both dental physicians were well-known researchers in their field. We were shown a few slides using a rudimentary projector that required inserting one transparency at a time.

Then came our turn. None of our delegates had come prepared to present anything that would be relevant to the audience we were facing. One smart guy in our group volunteered to talk about asbestos, which Chain Robbins, our delegation leader, politely turned down as not relevant. Finally, the delegate from the University of Pittsburgh thought of a topic that saved the day. Professor Dietrich Weyel offered to talk about lead poisoning among children. Children's teeth played an important part in Dietrich's presentation. The fact that we were industrial hygienists, not dental hygienists, did not seem to matter anymore. We were talking dental. That made our hosts happy.

Back at the hotel, we wondered if the visit was relevant to our IH mission. I don't think we advanced our knowledge of Russian hygiene, but it was a successful visit. I still fondly remember the wonderful reception and hospitality and above all the faces of the beautiful little Russian children, laughing, jumping, and screaming just like kids anywhere in the world.

A quick industrial hygiene lesson at the Hilton bar

A couple years later, I was having a beer at the Las Vegas Hilton, a sprawling property with more than a thousand hotel rooms. I was staying at the Hilton because there was no other place to stay within miles of the facility where Herbert (Herb) Mossner and I were conducting an exposure monitoring program for a power company client. The project site did not have a hotel or restaurant within miles. The closest place to find any nourishment in the vast Nevada desert was a little store run by a Navajo Indian family. The store did not carry a lot of merchandise. The inventory contained basically soda pop, beer, and lots of potato and corn chips. Soda pop and corn chips became our daily lunch for the duration of the project (cold beer was tempting in the hot Nevada sun, but we never consumed alcohol on the job).

The Hilton bar was not busy. Soon an attractive young lady walked in, sized up the environment, and decided to sit next to me even though there were many empty seats scattered throughout the bar. This was a wise decision. It was obvious she wanted to talk.

"Are you from out of town?" she asked, but before I could answer, she corrected herself.

"I suppose that was a dumb question. Most people around here are from out of town. The fact that that you staying at a hotel should have been my first clue that you are not from around here."

"Yes, I am from out of town. I am from Detroit, Michigan. Have you ever been to Detroit?" I asked.

"I feel sorry for you. Yes, I have been to Detroit. I don't have any desire to go back, but I am sure it is a good place for business."

"Yes, it is a good place for business. That is why smart people like Henry Ford built a big business empire in Detroit."

"My name is Joanna. May I know your name, and what do you do?" Joanna doubled up on her questions to get to the more important topics on her mind.

"My name is Jas and I am not a gambler, although many times I think maybe I am, but I will not bore you with that. I am an industrial hygienist and I do not expect you to know what kind of an animal that is."

Joanna laughed and said: "Then educate me on what kind of an animal I am looking at. I am all ears."

I was elated. Maybe my first impression of Joanna had been wrong. Maybe she was a student studying science or engineering at the University of Nevada, Las Vegas. I knew Las Vegas offered more than just slot machines. I started thinking, which can be dangerous in Las Vegas bars, here is a bright young college student who comes here just to relax a bit and to be away from her studies for a few minutes. Man! She could be interested in industrial hygiene if I could motivate her and explain the discipline. Perhaps I could steer her toward a master's degree in hygiene or ergonomics at Michigan or Berkeley.

"So, what is this hygiene thing that you do, and is this a nice profession? And does it pay well?" This time she tripled up on her questions. Perhaps I was going too slow, too deliberately. She did not have the whole evening for me.

I started describing what an IH does and what you have to study to become one and the accreditation process to become a certified hygienist.

She interrupted me again. I was taking too long. A short description would have been adequate. Exposure limits, statistics, and standard deviations were not of interest to Joanna, although I could have been wrong about the standard deviation part.

"Do industrial hygienists get paid well? May I know what kind of fee you charge for the job you are doing here in the Nevada desert?"

I hesitated. I don't throw around my billing rates unless there is a specific need. But in my effort to encourage the young lady to consider an IH career, I decided to share the information.

"My rate is $150 per hour plus travel expenses, including airfare, hotel, and food—and sometimes beer—but only if I am entertaining a client."

"That is wonderful. This means you can buy me the beer and charge it to your company. I could be your client. I am sure you use subcontractors that provide valuable services. Let me tell you, I provide equally useful services that you provide as a hygienist, and isn't it interesting that my rate is also $150 per hour? And I don't bill clients for airfare and only occasionally for the hotel."

"Yes, it is interesting. I am sure you deserve every penny of whatever service you provide, although I am not sure what service you provide."

"Don't play dumb, Jas," Joanna admonished me. "You do know what business I am in. You told me industrial hygienists are some of the smartest people. Do you want me to describe my professional services in detail?"

"Oh, no. It is not necessary. I can guess. We industrial hygienists are very perceptive. We deal with people. We deal with people's problems and strive to make their lives healthier and happier."

"Then you can make both our lives happier and healthi—" She hesitated at "healthier."

Joanna was not going to let it go. She wanted to know what I was going to do after I finished my beer. I told her I needed time to prepare for my visit to the client site where Herbert and I anticipated a very busy day. She suggested we meet at the bar at the same time tomorrow. I said I would be there.

It was a false promise. I had no intention of meeting Joanna the next day, but I wanted the evening to end amicably without insulting her. My mother had taught me to be nice to ladies.

I was still hoping that Joanna would consider becoming an industrial hygienist. She had the intelligence and a passion for people.

What a profession!

I love my profession even though I can never adequately explain to my friends and neighbors what I do that allows me to live in a nice place like Hawaii. I know that what I do touches lives even in small ways.

Talking about small ways, I will never forget an incident that Elizabeth K., a young ergonomist, once shared with me. Soon after graduating from college, Elizabeth worked for me as an ergonomist in Los Angeles, California.

One day Elizabeth came back all excited from a survey at an auto plant. She had met a sixty-three-year-old pattern maker who complained of pain in his wrist and his fingers. The cause was the repeated use of a thin pencil like a marker that the man used all day to draw his patterns for molds. The "pinch grip" he used to hold the marker exerted so much stress on the old man's fingers that he would go home with pain in his hand that kept him awake at night. Elizabeth instantly recognized the problem. She asked the man to give her the marker he used to trace his patterns. She then grabbed a roll of masking tape and started wrapping the tape around the marker until the barrel of the pen was an inch think. Elizabeth advised the man to use the "tape-fortified marker" from then on. She explained to the man that the crude alteration meant that now he would be using stronger parts of his muscles to grip the pen, and the forces he exerted would be distributed over a wider area and not be concentrated around his forefinger and his thumb.

Two days later, after Elizabeth returned to the plant, she saw a broad smile on the man's face from a distance.

"How did it work?" Elizabeth asked eagerly.

Before he said anything, the man moved closer to Elizabeth and engulfed her in an affectionate bear hug. He was almost in tears.

"You made my life worth living again. Last night I felt no pain. I slept like a baby, thanks to you."

Elizabeth assured him that she would try to find him a more permanent ergonomically designed marker.

Back at the office she sounded very excited. "Jas, I have had thanks from clients before. And I have had compliments on the jobs I have done, but I have never been hugged by a client for making his job more comfortable and safe. It is so sweet!"

From a dusty village in Northern India to the Mile High City, USA

I hope I can be forgiven for singing the praises of my profession. I cannot help myself.

As previously reported, I was born in a dusty village in Northern India called Gurne. My family was not wealthy. I had no rich uncles who sent me abroad or helped me find a job. I came to America with fifty US dollars in my worn and baggy pant pockets.

Luckily I discovered industrial hygiene. I had not even heard the word before. Industrial hygiene opened many doors for me. It brought me in contact with some of the brightest, friendliest, and most compassionate people that inhabit this planet. Industrial hygiene helped me realize the American Dream.

In 2005, the annual Professional Conference of Industrial Hygiene (PCIH) was held in the Mile High City (Denver, Colorado). That year I was selected to receive the prestigious "Henry Smyth Award," an honor that the Academy of Industrial Hygiene (AIH) bestows upon an IH professional who has "made significant contributions to the profession." The recipient of the award had to address the entire conference on a Tuesday morning.

Tuesday was my day. This was what I had waited for all my professional life. It was the biggest reward of all for me. Pay raises and promotions are nice, but nothing can trump peer recognition. I looked at the names of the people who had received this award before me. I was humbled. I promised myself that I would be sure to acknowledge my mentors, the people that had made this day possible for me. On the stage I would be composed. I would not get sentimental.

On the Monday night before the award ceremony, I could not sleep. I kept thinking, *How could I not get emotional tomorrow?* Just by chance I had stumbled upon a profession I did not even know existed, a profession whose only purpose was to protect lives, one that gave me a good life and so many opportunities.

Lying awake, I reflected on my life. I thought of the famous words of the Russian comedian Smirnoff, who in admiration of his newly found country—the USA—used to say:
"What a country!!"

I wanted to modify Smirnoff's slogan. "What a profession!" I blurted out without realizing that I awakened my wife in the process.

"Go to sleep. You have a big day tomorrow," my wife advised.

It was easier said than done. *What a profession!*

CHAPTER 5

TERRE HAUTE, INDIANA –
THE VACATION HAVEN

Herb Mossner always wanted to spend his vacations in Terre Haute, Indiana. It is not that he did not know the world beyond Terre Haute. As an accomplished Board Certified Industrial Hygienist (CIH) he had been to the Tinsel Town (Los Angeles) a dozen times, the Gambler's Haven (Las Vegas), and places like San Francisco. But according to him, those were the kinds of places the boss forced you to go to for work. If you wanted a real vacation, you went to Terre Haute, Indiana.

Strangely, this was not the first time I had heard someone describe Indiana as the place to go if you wanted to have the time of your life. The comedian Steve Martin also extolled the virtues of Terre Haute, Indiana, as the ultimate fun capital. Steve Martin even recommended a few of the establishments in town to patronize if you were a fun-loving person. I did not wonder about the places in Terre Haute that Steve Martin recommended. After all, I was watching a comedian whose job it was to make fun of people and places.

But Herb's devotion to Terre Haute intrigued me. He was not joking. I tried to find out what was there that was so special that had been kept a secret from the rest of us less-informed souls. Herb would not tell me. I did know that his mother was there. Being close to his mother, I could see his desire to see his mom, but to build up Terre Haute as a vacation destination? That was too crazy even for Herbert Mossner. As I got a little closer to him,

I did start understanding his reasons for spending vacations in his beloved town, but we will deal with that later.

In 1978 I joined a company in Detroit, Michigan, that specialized in industrial hygiene and safety. Herb was already there. You could say he inherited me. Initially I was puzzled, intrigued, and even sorry for his inheritance. I could never figure out where this guy stood and where he was coming from. Everything he said to me had to be passed through filters and decoders to get the straight meaning. His verbal communication was always loaded with puns, humor, or sarcasm.

Time can heal wounds. I started liking the man. His sense of humor was delivered in a mature literary way and always in a measured but blunt style. I was getting to like his humor to the point that often I would coax him to come up with his weird pronouncements while on the surface I kept reminding him to watch his tongue. After all, I was his supervisor. I had to act responsibly.

Herb was a talented writer. I would challenge anyone to find a mistake in a report he had authored. Not a comma, a hyphen, or a chemical fact would be missing. Perhaps this was a result of his education at Purdue University in West Lafayette, Indiana, where he graduated with a degree in chemistry. I always showed Herbert's reports as a template to other staff to show what a consultant's report should read like. The only problem was finding a completed report. Herb would procrastinate for weeks while I, as the department manager, got complaints and agitated phone calls from clients whose reports were overdue. On a few occasions my frustration would reach the boiling point, resulting in unwanted tension between us. Sooner or later this always ended amicably. We both needed each other. I needed him more. His tardiness notwithstanding, he was the best technical professional in the group—a pain in the rear, but a certified (CIH) one. This forced admiration resulted in something of a bonding between us. I never conveyed this to him while we were colleagues. I was afraid of what he would say if I mention the word bonding.

What's in a name?

Herbert was proud of his German name, and why not? He told me that many prominent Germans belonged to the "Mossner" clan. To make it even more special, his name was spelled with "double S," when many of his less-fortunate namesakes (the Mosners) had only "one S." The significance of that distinction was totally lost on me, but it seemed terribly important to Herb Mossner.

In 1981, the annual conference of the American Industrial Hygiene Association (AIHA) was in Portland, Oregon. Each year the company in Michigan where Herbert and I worked would send a large contingent to this conference because industrial hygiene was our "bread and butter." The Portland conference was no exception. Several of us, including Herbert, eagerly headed to the event in which we had been waiting for with much anticipation for a whole year. We were going to stay at the fancy Portland Hilton, the designated conference headquarters. In our eagerness, we arrived there one day early without having confirmed hotel reservations for the first night. Upon arrival at the airport, we found out that no rooms were available in the city. Our best option was to stay at a motel near the airport. Even at the airport motel, we had to double up because not enough rooms were available for everyone in our party.

As luck—if you can call it luck—would have it, Herb and I were assigned to share a room. That was bad news. I prepared myself for the pending disaster. I had heard stories, mostly from Herb himself, that he had a difficult time sleeping. He had to have a TV running while he slept. Moreover, he snored so hard that he awakened himself with his own noise, a phenomenon he described as self-inflicted torture. I mentally prepared myself for all this, not knowing the horrors that were yet to unfold.

The evening started pleasantly enough. We settled down and planned to go to dinner at a nearby restaurant. Soon I saw him furiously shuffling through the local phone book, jotting down numbers and doing some math. He would pause and recount to make sure he had the correct numbers.

"What in the world are you doing? What is so fascinating with the phone directory?" I asked.

"I am counting the number of Mossners listed in the phone book. The reason it is taking me so long is because I must divide all the Mossners into two categories: the ones that have only one S in their name vs. those lucky ones, like me, who are blessed with two S's in our name."

"You are being ridiculous, Herbert. I think both names, whether with one S or two, are decent names. Right now, in my opinion, you are not acting like either kind of Mossner. You are behaving like a half-S Mossner."

It took him a few seconds to realize what I had said. "Oh man, that is a low blow. That's not funny. You think you are funny? You are not," he declared.

I still think it was funny.

Odor eaters

Herb was not yet ready to go to dinner. He carefully pulled out a package that he had bought at the local store before getting on the plane. He then unpacked some rubber pads shaped like feet. His package included a pair of scissors and a marking pen. Next, he stepped on one of the pads and, with the help of his marker pen, traced the geographical boundaries of his left foot. He repeated the exercise for his right foot, which he told me was a bit different from his left foot.

"What in the world is going on? What are you doing, Herbert?" I was dying of curiosity.

"I am cutting my odor eaters to size," he replied calmly. "These ones are the best ones money can buy. The pads are impregnated with the purest activated carbon, the kind with the largest surface area that you can get only from charcoal made from burnt coconut shells. These babies can eat foot odors like those microorganisms in a well-fed activated sludge sewage treatment plant." He continued, "I know you are hungry, but just wait for another few minutes until I check the precise fit, and then we can go for dinner. You will thank me that I did this before dinner."

The comparison of Herb's foot odor control to a sewage treatment plant just before going to dinner helped curb my appetite quite a bit. But I agreed. It was better to do this before dinner rather than after.

Herb compensated for his unappetizing acts before dinner by telling me some funny stories – including one about the time he struggled to figure out the gender of a New York City advertising consultant who was assigned to work with him on a promotional video.

That night my sleep was interrupted several times by his massive noise-making machine, his nose. He apologized for it every time he woke me up. One time the noise and accompanying vibrations in his throat were so loud that I shrieked. I was sure that a truck had crashed into our room because our motel was near a busy highway.

The next day, thankfully, we were able to claim our reserved, single occupancy rooms at the comfortable Portland Hilton.

Summers in Terre Haute

Sit on the porch, drink beer, sweat, and stink

Summer was approaching. Herb put in his request early to take his vacation and go to his beloved Terre Haute for some R&R during the peak of summer. The company in Detroit where we worked was short on manpower because the workload was heavier than it had been in years. It required tight planning to keep client commitments. As a consulting company, we could not afford to default on our obligations. We were a service company. When a client said jump, we would jump. Sometimes we jumped higher than necessary.

Only a week before Herb was to go on his vacation, a large construction company based in Idaho came to us with an urgent project. They were building an addition to an existing multistory power plant in the Nevada desert. Construction had been going great until the new structure reached the third story right next to the existing power plant building. The steady and reliable desert wind would occasionally change direction. Combined with the complex meteorology of the desert, the wind would funnel acrid sulfur fumes from the existing smoke stacks toward the scores of construction workers at the third level of the new structure. Experts said it was "inversion." Some called it "building down wash." Whatever the phenomenon, it sent the workers gasping for breath, and they would head down to the ground to escape the noxious onslaught. The main fear was that it could cause all kinds of maladies and health damage. As the work reached higher elevations, the problem got worse. Finally the worker's union told the construction company that their members would walk off the job if the company did not promptly fix the problem. This would cause an extended project shutdown, millions of dollars in damages, claims, and who knows what else?

The construction company called us for help. We had to send someone to Las Vegas (the construction site was a forty-five-minute drive from Las Vegas) to study the situation and return with solutions as soon as possible. One suggested solution was to install a gas monitoring system with alarms and buzzers to warn the workers to head for the ground level before the noxious gasses gagged them. The control system had to be reliable.

I decided Herb Mossner was the man to do this. He had prior experience with this type of situation. He could argue with the union if necessary and articulate the soundness of his approach.

I wondered how to tell him to postpone his Terre Haute vacation. He had been dreaming about it for weeks. But it had to be done. I walked up to his desk and apologized but asked him to postpone his vacation for a few weeks.

He was stunned. He could not believe what he had just heard. He'd had those dates blocked out on his calendar for months. How could anyone do this to him? He got out of his chair, moved a little closer to me, gave me the most disgusting look, and yelled: "I knew it! I knew it! I knew that you were going to do this to me. You always do this, Jas (I do not). Like an idiot, I believed you. I should not have."

The ranting went on for a while. Finally, he calmed down a bit and said, "You know damn well that I will do this for you, but you owe me one."

"Yes, Herbert, I owe you one. You want it in writing?"

"No, I don't. How do I know if your written warranties are worth anything?"

I went back to my office. I was as upset as Herb was. I had reneged on my promise to the man. He had blocked the dates well in advance. How would he ever trust me again? My credibility was everything to me. I did not want it destroyed. I decided I would tell him he could go on his Indiana vacation as planned. I would figure out something. Maybe I would go to Nevada myself despite many other things on my plate, and knowing full well that I did not have the knowledge Herbert had on the subject.

I went back to his office and started apologizing for wrecking his plans and started to tell him that he could go to Terre Haute as planned. I said to him: "I am sorry, Herbert. I hope I did not mess up your plans to do all the special things you had planned in Terre Haute."

He did not let me complete the sentence. He had cooled off by then. He said, "It is okay. I will manage it. I had not planned anything that cannot be duplicated. I was just going to go home, sit on the back porch, drink beer, sweat, and stink."

I could not let that go. "You can still do all that. The hot and humid Indiana summer is still young. You will have several more weeks to sit on your back porch, drink beer, sweat, and stink," I assured him.

"You are always under this misconception that you are funny," he reminded me with all seriousness.

Herb went to Nevada, gathered the information, took copious notes, and gave the information to Jerry, the engineer at our company, to start working on a system while Herbert vacationed in Indiana. Everything seemed to be going well.

While Herb was vacationing in Terre Haute, I received a call from a vice president at the Idaho construction company. An urgent meeting would be held in Las Vegas to explain the strategy we had proposed to the rank and file of the Iron Workers Union. We had to explain how the construction company would protect union members from being gassed by the sulfur fumes. The VP proposed to me that Mr. Mossner and I make a detailed presentation to the union in a big hall the company had rented for the occasion in downtown Las Vegas. We needed to be prepared to answer any and all questions, remembering that some of those questions could be pointed and even hostile. Because the issues involved were sensitive and volatile, the construction company suggested that we should first come to Boise, Idaho, to discuss with the company executives what our approach would be. I needed

Herb to go with me, but I was not going to disrupt his heavenly Indiana experience. Memories of our friendly clash were still vivid to me.

I decided to go alone. I could handle this. I was being paid for such emergencies even if it included putting up with insults, humiliation, and rebuke. This is one of the prices you pay being a boss.

The Idaho executives were disappointed to know that Mr. Mossner would not come, but I assured them not to worry. I had the information. We started our discussions. Everything was going great except when they needed a certain detail on the gas monitoring plan we were designing. Unfortunately I could not tell them the needed details.

"Call 'Herby,'" the vice president said as he pushed the telephone in the middle of the mahogany conference table toward me. He suggested that I put the phone on speaker so everyone around the table could hear and ask questions.

I did not have Herb's phone number with me. I called my Michigan office to find his home number in Indiana. I put the phone on the speaker and dialed the number.

"Hello," a female voice answered.

"May I please speak with Herbert Mossner?" I was very conscious of interfering with his dream vacation.

"Who shall I say is calling?" the woman at the other end inquired.

I knew it was his mother. He had talked about his mother on several occasions. He was very fond of his mother, which would explain (to some extent) the pull he had for Terre Haute.

"Mrs. Mossner, this is Jas Singh from work. I am extremely sorry to disturb Herbert from whatever he is doing after having promised him I would not bother him on his vacation. It is urgent that I talk with him for a few minutes. I hope I am not disturbing him from something important."

His mother seemed to know who I was. Apparently Herb had mentioned me to her, which made me nervous. She was very gracious and reassuring. She said: "Don't worry, Jas. He is not doing anything important. He is just sitting on the back porch, drinking beer and sweating. It is hot and humid here." She did not say "stinking," which was among the joys of vacationing in Terre Haute in summer as Herb had expressed to me.

I suddenly became more nervous. All this conversation was being heard by everyone around the table loudly and clearly. I could not change anything at this time. Herb picked up the phone and yelled, "Hello, Jas."

Before another syllable could come out of his mouth, I shouted, "Hey Herbert, I am sitting here with several executives of the Idaho construction company and we have you on a speaker phone, and we have a question that only you would know." I continued . . . without giving Herb the chance to break in and say something outrageous.

The VP of the company came to my aid by chiming in, "Mr. Mossner, we appreciate the work you have done putting the plan together on such a short notice. Our engineers think it is great, but we have a few questions."

Herbert was meek as a lamb. He answered the questions to our client's satisfaction. He was pleasant and courteous, the kind of courtesy and politeness he never bestowed on me in the office. Nevertheless, I was delighted. I rather he displayed this courtesy to a client than internally.

The company executives were satisfied. The VP apologized to Herb for disturbing his vacation but could not resist saying: "Thank you, Mr. Mossner. You can go back now to your porch and resume your vacation. I assume that it is a hot summer in Indiana. Please have another beer on our behalf."

I turned off the phone and everyone burst out in laughter.

A couple weeks later Herb returned from vacation but immediately started repacking to go back to Nevada to oversee the implementation of the smoke control plan. The union seemed satisfied with the monitoring strategy and the alarm systems in place to warn them of pending fume concentrations.

I still see Herb from time to time, usually at IH or safety conferences. We talk about the old times. Soon a crowd gathers, composed mostly of the people we both know. It becomes a laugh riot. Nostalgia takes over.

I love it!

CHAPTER 6

HOTEL CIDADE DE DAMAN SEASIDE

Practicing Industrial Hygiene in Portuguese India

Portuguese explorer Vasco da Gama landed in Calicut, India, on May 20, 1498. Anchored off the Malabar Coast, he invited native fishermen on board his vessel and bought some local items from them. The Portuguese conquistador, however, had much more on his mind than souvenir trading. Over the objections of the local Arab merchants, Vasco da Gama managed to secure a letter of concession for trading rights from the area's Hindu ruler. Later on, local Calicut officials decided to temporarily detain Gama's Portuguese agents as a security for payment. This annoyed the conquistador, who decided to abduct several of the natives and sixteen fishermen as a show of his displeasure. Thus started a sustained campaign by the Portuguese to annex large territories of India, establishing Portuguese colonies all over Southern India (see map). The Portuguese gave the name "Estado da India" (Portuguese India) to their prized possession. Estado da India was established in 1505 as a viceroyalty of the Kingdom of Portugal. Bombay (now called Mumbai) was part of Portuguese India at that time but was later given to Britain in 1661 as part of the (Portuguese) Princess Catherine of Braganza's dowry to Charles II of England.

At the time of India's independence from Britain in 1947, Portuguese India included a number of enclaves on India's west coast, including Goa proper, as well as the coastal enclaves of Daman (Damão in Portuguese) and Diu. The Colony of Goa had authority over all Portuguese possessions in the Indian Ocean, from southern Africa to Southeast Asia. Goa became the capital of Portuguese India until its merger with the Indian Union in 1961.

Cidade de Daman

Portuguese footprints reach all over the western and eastern coasts of India. Daman was incorporated into the Republic of India in December 1961 after a battle between the Portuguese and the Indian Army. The battle left four Indians dead and fourteen wounded. Portugal suffered ten dead and two wounded.

Daman is a multilingual city. Languages spoken include Portuguese, Konkani, Gujarati, Marathi, Malayalam, and others. Located about 120 miles north of Mumbai, Daman, along with the neighboring towns of Vapi, Bhilad-Sarigam, Bilimora, and Silvassa, has become an important manufacturing hub. A variety of products—including pharmaceuticals, fertilizers, chemical dyes, toys, printing inks, plastics, and electronic components—are manufactured near Daman.

I was always fascinated with Goa, Daman, and Diu. When I received a request to conduct an industrial hygiene project in Daman, I could not contain my excitement. Although I was born in India, I had spent the first one-third of my life in the north. India is a vast country and transportation was difficult and expensive so I had no idea what Estado da India was like. Having grown up in the hot and arid north Indian plains, I dreamed of the lush and fertile lands within India where people spoke foreign languages, especially European languages. I had even heard of a town in India where people spoke Danish. The town was called Fort Dansborg. At one time Fort Dansborg was the capital of Danish India. Danish India had included several Danish towns and the Nicobar Island. Danish India was later sold by the Portuguese to the British in 1845.

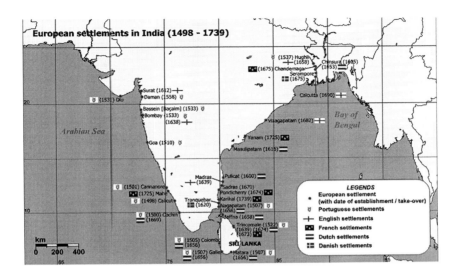

When I thought of Goa, Daman, and Diu, I imagined lush green landscapes where coconuts and bananas grew wild and the palm trees swayed from the trade winds. Several times I came close to realizing my dream of visiting the three Portuguese enclaves. Up to that point, I had never made it there. One small problem had been money, or a lack thereof!

My wish comes true

Finally, more than forty years after leaving India, my wish had come true. A pharmaceutical company in Daman requested that we conduct a survey to determine the extent of exposure to their workers by suspected toxic chemicals they were handling. This was my calling.

It may not be widely known, but India has become a pharmaceutical manufacturing giant in recent years, ranking among the top pharmaceutical producers in the world. Because of my chemistry background I liked practicing industrial hygiene and safety in the chemical industry, the pharmaceutical industry in particular. My PhD was a curious mixture of physical and inorganic chemistry, a sort of a hybrid that reminded me of the mixed

German Shepherd and Cocker Spaniel dog we once owned. I liked the pharmaceutical work because of the extremely toxic nature of some of the ingredients, known in the industry as the active pharmaceutical ingredients (API). Many people are surprised when you mention that the APIs responsible for the beneficial medicinal effects are highly toxic themselves. When you think about it, however, a therapeutic agent would logically be expected to be "active" (toxic); otherwise it may have little biological potency.

The key to chemical safety is the dose. In fact, the whole field of occupational hygiene comes down to identification and control of the dose. This is best said by philosopher Paracelsus who gave us the definition of the word 'dose' almost five hundred years ago.

"All substances are poisons. There is none that is not a poison. The dose differentiates between the poison and the remedy."

Technology, Bharata Natyam, and Veena

When the opportunity to do some IH work in Portuguese India came about, I immediately called Chitra Murali, an industrial hygienist in Delhi. Chitra, a graduate of the Indian Institute of Technology (IIT) in Bombay, is a multi-skilled individual. Her skills are not limited to environmental engineering, the discipline she studied at the IIT. Through work experience Chitra developed an interest in the field of industrial hygiene, no doubt because of constant propaganda and brainwashing from me and other company hygienists.

Chitra is also a Bharata Natyam enthusiast (a classical Indian dance form) and an accomplished classical Veena virtuoso.

Chita's sense of humor is one of her many assets. This quality is especially useful when traveling long distances for extended periods in unfamiliar places in the company of technical people. Such travel is a part of life for most IH professionals.

Beachfront luxury

To travel to Portuguese India, I first flew from Honolulu to Mumbai. After a long taxi ride, Chitra and I arrived in the city of Daman. At first, it did not look any different than the rest of India. Once we got to Hotel Cidade De Daman (City of Daman), however, things started looking up. Our Daman host had informed us that all of the newer and nicer hotels were filled up, as the area was experiencing an industrial boom like many other parts of India. As a result they booked us at an older hotel called Hotel Cidade De Daman.

Hotel Cidade De Daman was the kind of place where I had hoped to stay. We didn't care that it was not modern and plush like the newer, fancier hotels in the area. Hotel Cidade De Daman was situated right on the oceanfront. Unfortunately, it had no beach. The tides were so strong and frequent that it became a guessing game as to when and where the beach would appear if it existed at all. The constantly changing seascape kept us guessing and interested.

Staring is fun

After getting our hotel room assignments, we instructed the hotel bell boy to deposit our bags in our rooms while Chitra and I dashed straight to the seaside restaurant before it closed. A young and eager waiter named Aziz promptly appeared, greeted us, and assigned us the "best and the finest" table closest to the beach. He then parked himself a few feet away to stare at us from a safe distance.

Curiosity was killing him. He was probably thinking: *Who is this man who is quite a bit older than the pretty lady with him? They do not seem to be related and neither of them seems as though they belong to this place. Something is surely amiss!*

Unlike many other places, staring is not necessarily an impolite or threatening gesture in India. It is more like a hobby, and it simply shows curiosity and a secret desire to get to know you. When I travel to India I get stared at all the time, which is surprising since I look no different than the hundreds of millions of other people there. During my most recent visit, my wife (who also gets her full share of stares when travelling with me in India) explained the reason to me. People were staring because I always wore flowered Hawaiian shirts, comfortable attire to wear in any warm tropical country. Flowered shirts, however, look very foreign in India. To look educated and respectable in India, it is customary that you wear heavily starched white shirts with a suitable tint of Indigo. Knowing this, I purposely wore flowered shirts to draw stares. I welcome stares because it means that someone finds me interesting.

Casanova de Daman

Enough staring! Aziz finally gathered enough courage to ask us some very personal questions. He turned to me and said: "Welcome, sir. We are so glad to have you good people visit our fine hotel." Then he turned about 15 degrees left toward Chitra while still addressing me and asked: "Is the young lady your daughter, sir?"

Chitra winked. I searched for an answer and said: "No, Aziz; she is my girlfriend."

Aziz stumbled and almost dumped a bowl of the steaming Mulligatawny soup in Chitra's lap. He stared at us like a deer into oncoming headlights at night. Just to be sure, he looked away to refocus and recalibrate his vision, stared again, and said:

"Well, sir, you must be a very rich man. I kind of guessed that by looking at you."

"Why do you say that? I am not rich at all. You don't think I can attract a pretty young lady with my charm alone?" I challenged him.

"Oh, no sir, sorry sir, I did not mean to imply that. I am sure there are hundreds of pretty ladies in Daman who would not mind too much having dinner with you, sir."

"You just restored my confidence and self-esteem, Aziz." I thanked him.

"Anytime, sir," Aziz replied as he returned to the kitchen to bring us some fresh bread.

Suddenly a huge monkey appeared out of nowhere and lodged

himself squarely on the sea wall about ten feet away from us. Although the gender of the primate was uncertain, I guessed it was a male because of his size. The primate paid no attention to Chitra but, to my discomfort, focused squarely on me while completely ignoring the waiters who were trying to shoo him away. Soon I realized the reason. He did not like my dining etiquette. What kind of an idiot would eat spicy Goan curry with a fork?

Realizing my stupidity, I dropped the fork and dove into the delicious gravy with my bare hands. The gesture seemed to please the primate. Short of clapping hands, he gave every sign of approval.

A smaller primate then arrived on the scene, walked up to the big monkey, and appeared to whisper something to him. I was sure it was a female.

Both of them bared their teeth, displaying the classic monkey grin, then left. I wondered what the lady monkey whispered to the big fellow.

Aziz came back with steaming bread. Relieved that the animals had gone, he tried to resume the discourse. He had to know how Chitra and I were connected.

Finally I told him that we were coworkers in town to do an industrial hygiene survey. The word industrial hygiene did not ring a bell with him, but he felt compelled to say something.

"I know the word hygiene, and looking at you two I could have guessed you were the hygiene kind of people," he said.

Emboldened by the fact that we were merely coworkers, he leaned toward Chitra and asked for her 'mobile' (cell phone number). Chitra struggled to decline his request as politely as possible, but she failed.

"Sorry, Aziz, I do not give my private phone number to a waiter every time I go to a restaurant."

Disappointment was evident on Aziz's face, but he persisted. Okay, fine, then can you please give me your work number?" he asked.

Chitra passed the buck (one of her specialties) and said, "Jas will give you his number. It is the same number for me as well." Aziz was disappointed but found this to be a reasonable alternative. He was thinking that even my number could be of value. Perhaps I could sponsor him for a permanent visa to the United States.

Neither Chitra nor I parted with any phone numbers, but we left on good terms with Aziz. We promised to return to one of the best exotic hotels on our next stay.

The project in this tiny Portuguese enclave in India reaffirmed my belief that there is no such thing as a boring industrial hygiene assignment. Industrial hygiene is a people profession. Everything is about people; even sampling devices are attached to people (usually their shirt collars) and not to metal stands, furniture, or walls.

CHAPTER 7

FOREVER YOUNG

Professor BK Nor is not really a professor, but he acts like one. He is passionate about teaching his subject—personal protective equipment (PPE)—to anyone who will listen. In some ways the professor is passionate about a lot of things, but we will not go into that quite yet.

BK is an accomplished health and safety professional in Malaysia. He graduated from one of the top universities in the United States. PPEs, the kind of devices used to protect people against exposure to toxic substances in the workplace, are his area of expertise and his business. In my opinion, he is the best in the field in his native land.

For Westerners, especially Americans, Malaysia remains somewhat of a mystery when interacting at a social level. That is a pity. They are missing out on a lot of fun, camaraderie, and most of all, belly laughs. To discover the true friendliness, warmth, and robustness of the Malaysian culture, they need to meet Professor BK Nor.

BK may very well be the friendliest man in Southeast Asia. I challenge anyone to make BK angry. It is not possible. If you say something weird or even something offensive, two things will happen: The professor might stare at you in surprise as if you are an alien species unfamiliar with normal courtesies and respect, but being alien, you should be forgiven. In such situations, he will smile a little and say, "That is interesting." If, however, you say something funny—it matters not whether it is off color—BK will laugh like a wild hyena. If you were sitting in the fancy lobby of a five-star hotel, like we were in 2012 in the Singapore Fairmont Hotel lobby during the American Industrial Hygiene Association (AIHA) Asia Pac conference, he would make certain that the entire lobby shared his laughter. He would then extend his hand to you for a warm handshake.

Consider yourself lucky. You have just gained entry into the warm, friendly, and hospitable Malaysian society and will forget about the more formal and reserved Singapore style. You would instantaneously know that this is a different level of social interaction. I don't want to imply that Singaporeans are not friendly. I have many wonderful and hospitable friends in Singapore, but even they agree with me that it takes more effort to get to know a Singaporean. For example, try getting the attention of a casual Singapore citizen on the street. You cannot do this even if you put on a clown suit and beat on gongs in the middle of Orchard Road. On the other hand, if you own the very latest electronic gadget—maybe the Apple smart phone generation X—you may have better luck.

Better living through chemistry

BK owns his own business. He is a supplier of PPE. To a non-technical person this may mean safety boots, hard hats, gloves, goggles, ear plugs, dust masks, etc. These familiar and not very technical or complex devices are the kind of things you can pick up at a Home Depot or Lowes or any mom and pop hardware store. In reality, PPE is much more involved and requires real expertise and decision making, especially when it comes to selecting, testing, and maintaining respirators (aka, masks) that protect workers from breathing hazardous chemicals. BK excels at this more technical side of the PPE industry. As soon as the subject of respirators comes up, the professor lights up like a 150-watt incandescent lamp. He talks about his education at the University of Cincinnati; the PPE seminars he has attended at prestigious American schools like Berkeley, Michigan, and the Lawrence Berkeley labs; as well his conversations with some of the PPE gurus in the west. Then he abruptly changes the subject, and with a nostalgic sigh, he will turn to you and say: "Oh man, that Cincinnati is a fun place."

That always cracks me up. "Cincinnati is a fun place! Do you mean, Cincinnati, Ohio? Are you kidding me? You are not smoking anything funny, are you? I don't need to tell you, man, that smoking Da Da (narcotics) in Malaysia can get you a life sentence."

This will guarantee a hyena laugh and, when recovered, he will say: "No, I have not been smoking anything. You know I don't even touch a cigarette. I am a health professional. I am a health nut, as you well know."

Do I know that!

Pharmacy in an e-Bag

Professor BK is a walking pharmacy. You can confront him anyplace, anywhere, and ask him if he is carrying any medicines with him.

"What for?" he will ask and then volunteer to show you everything he has in his e-bag pharmacy. Before you say anything, out comes the whole pharmaceutical paraphernalia. Colorful bottles and packets emerge one by one from the e-bag like a magician pulling out live doves from a black velvet bag.

One day at the lunch table of the biannual conference of the International Occupational Hygiene Association in Kuala Lumpur, I mentioned BK's e-bag pharmacy to John Henshaw, the former head of the US Occupational Safety and Health Administration and the keynote speaker at the conference. BK, who was sitting next to John, seized the moment and started pulling out his miraculous potions one by one and arranging them carefully on the lunch table until there was no more room on the table. He had forty-three health elixirs in all. These included drugs prescribed by his physician and also drugs not necessarily procured with a physician's signed slip; it is common to find prescription drugs in Malaysia without a prescription, just as in Mexico. I guess a pharmacist could get in trouble for that, but an innocent buyer is never aware of this and may not even care if he/she is ill from bacterial infection and does not know where else to go.

A family soccer team

BK was on a roll. He told all of us around the lunch table that in addition to carrying his e-bag pharmacy with him wherever he went, he also received daily health alerts from a publication to which he subscribed. He then showed us the health bulletin of the day that he had just received on his smart phone. The health alert read: "Nature's Own Blood Pressure Regulators." The bulletin promised: "Sail through effortlessly into old age and rewrite the story of your life."

As if on instinct, BK turned toward me and said: "Dr. Jas, you claim to be a writer, so here is a topic for you. You can write my life story and, to make it even more interesting, I am going to tell you something about my life that may have escaped you earlier. I am a soccer enthusiast like most Malaysians, but unlike most Malaysians, I want to have my own soccer team someday. Mind you, my very own," he emphasized.

"What do you mean your own soccer team? Are you telling me you want to buy Manchester United or the Los Angeles Galaxy? Do you make that kind of money peddling PPE?"

"No, I do not, but I have another resource. I have seven children. Did you know that?"

Before I could respond, he broke out again in his patented hyena laugh, forcing everyone around to pay attention to us. When recovered, he continued, "You know that I am only four players short of a full team." He then raised four fingers of his left hand, separated equidistant from each other to make sure that even the visually impaired could get the count right. He continued, this time addressing both Henshaw and me saying: "You know I am healthy, I exercise, I take good medicines, and I heed the advice from my daily health bulletins. My beautiful wife is a few years younger than me and in fact is in her peak years of childbearing potential. I will ask her to allow me to complete the roster before the decade is over."

"Excellent plan," I clapped wildly. John smiled but did not react the way I did. He was the keynote speaker. He was obliged to show restraint and dignity.

On another afternoon in November of 2012 I was witness to a display of the e-bag pharmacopeia when four of us were having a drink in the lobby of the Fairmont Hotel. The question of BK's miracle drugs came up, and BK again graciously agreed to display the walking pharmacy to another distinguished visitor to the conference, neatly laying out on the coffee table all forty-three miracle drugs. It was such a spectacular display that we all reached for cameras, including the serving hostess. The pharmaceuticals included over-the-counter Chinese and Malay herbal remedies, many of those being youth- and stamina-enhancing elixirs. Among those were Chinese herbal wonders that could extend your love life well beyond what Western pharmaceutical companies had been able to produce.

One of BK's miracle potions was a pentagonal blue pill quite a bit bigger and bluer than the familiar product from a major pharmaceutical house. When I asked BK what that was for, he casually brushed me away. He smiled and simply said, "Oh Jas, this is not for you." I did not question him any further.

Marrying a chemist can be hazardous to your health

BK was not shy when talking about his personal life. In fact, he relished it. And why not? He was a successful man. I am not sure where he started, but where he is today is something to be proud of. He has many friends, a nice house, seven healthy children, a fat bank account (he needs one), and a trophy wife. Moreover, he is bestowed with an extremely good nature. He once told me he could marry just about any woman in his ethnic group. I wondered whether it was his money or his wit, but it mattered not. He had the proof. He showed us his new wife's picture, a beautiful and sophisticated woman about half BK's age.

Being always conscious of risk and reward considerations, I blurted out: "Be careful, BK. Have you considered the risks associated with your trophy acquisition?"

He gave me a curious look through his smiling dark eyes as if he had guessed what I was thinking. "What risks do I face in marrying a younger woman who seems to love me and the one I know half the men in KL (Kuala Lumpur) would kill to have?"

"That is precisely my point, BK. Think about it . . . She could get any of those successful, dashing younger men in town, and has the thought crossed your mind that for that reason she might try to poison you? And, my friend, if she wanted to do that, she would have no trouble finding some poison. All she has to do is to reach inside your own e-bag pharmacy."

"You are a wicked man, Jas." Another hyena laugh, and BK said: "You are absolutely right. If you want to know the truth, this has been on my mind for the last thirty years, even when I was married to my first wife. My first wife was also a very pretty woman, composed and kind and, to tell you the truth, far too good for me. I always felt I did not deserve her, so she might have wanted to poison me." BK continued, "Jas, you would get a kick out of this. She was a chemist just like you and a good one like you." I nodded in complete agreement. He continued, "I started thinking that she would poison me some day so I usually kept my medicines hidden, but I also did not have many of these drugs those days. You know, I was a lot younger and did not need all the supplements. The opportunity to get poisoned is a lot more real now for obvious reasons, and, by the way, being a PhD chemist, my friend, do you know any good antidotes you can recommend? I am thinking, what are your friends for after all?"

"Friends are here to keep friends from being poisoned by their spouses," I assured him.
"I can search for potential antidotes and try to see if you already have them in your e-bag. Actually, she would be better off to go outside for a poison with low LD 50 (lethal dose of a chemical that under controlled experimental conditions will result in killing 50% of the test animals). A more potent poison from the outside will be preferable because you don't need a whole lot to put somebody to sleep, plus you would not get suspicious if you found out that one of your own e-bag poisons was missing."

"You can also search for antidotes on your own starting with the *Toxic Chemical Encyclopedia*, and I can recommend to you an excellent book on the subject. It is called *Handbook of Poisoning* by Dr. Robert H. Dreisbach, MD, PhD, Clinical Professor of Environmental Health at the University of Washington, Seattle. I own this book and can bring it with me next time I come to Kuala Lumpur. Thinking more about it, though, you should not postpone this because it is a matter of life and death. As the English say: Don't be 'penny wise and pound foolish.'"

BK agreed wholeheartedly.

Boat to Batam
The conversations then shifted to the hotels where everyone was staying. I complained about the high hotel prices in Singapore. I told BK that I was staying at the Novotel Hotel near what is called Clark Quay, a popular area

along the water replete with restaurants and bars, including some fine American establishments like Hooters, known for their spicy chicken wings. I told BK that at my hotel I had access to the Novotel Club Lounge that had good food and hostesses who were polite, friendly, and sophisticated. But I told him that I did not recommend the lounge for him because then he would flash around his wealth and wit and would want to marry one of the hostesses, thus creating another opportunity to be poisoned.

Another hyena laugh and recovery, and then BK said: "No, my friend, I am done with marriages. Moreover, I cannot tolerate any more risk until you find me an antidote approved by a PhD chemist and a board-certified industrial hygienist."

BK suddenly changed the subject again and said: "So, what are you paying at the Novotel with the club lounge and hostesses and all that?"

"Well, you know, with the lounge and all that, Novotel is not any cheaper than the Fairmont, but at the lounge I essentially eat for free—not just breakfast, which is exquisite—but also dinners, which are mostly hors d'oeuvres but are still better than most street vendor food."

"Sounds good, but still you are wasting money. Okay, your company's money. You can take a boat to Batam, Indonesia, only forty-five minutes away, and you can have a decent room for fifty American dollars. The hostesses are equally friendly and, if I may say so, probably friendlier. It may not be as glitzy, but it provides all the comforts that only a home can provide."

"Are you suggesting I commute to Batam, Indonesia? Are you crazy? You mean I should go to another country every evening after work just to sleep in a cheaper hotel room? Is that what you are suggesting?"

"Precisely," BK said calmly. "Hundreds of people do that every day. On a typical evening, a statistically significant population of Singaporeans takes a boat to Batam, although most of them come back home to sleep. But an overnight stay is what I recommend. You will like it and will return home with most of your money still in your pocket and your body parts intact."

"That is very sound advice, BK, and after all what are friends for? Unfortunately, I cannot do that because according to my company's travel protocol, every time I go there I would have to file a risk assessment statement and health & safety management plan, which has to be approved by a top officer of the company."

"There is no more risk in Batam than any other place in America," BK shook his head in disagreement. "Every place has its own bad areas and bad characters; you guys are just wimps."

I agreed and we prepared to leave.

"And when will you be back in KL?" BK asked. "Let me know in advance and I will meet you wherever you will be, and if you will be teaching any of your industrial hygiene workshops, I will present a session on respiratory protection and will share some new information that I learned at the recent conference organized by the National Institute for Occupational Safety and Health in America."

We shook hands and BK left, still laughing like a hyena. I don't know what was so funny, but I could still hear the laugh long after he had cleared the revolving doors of the Fairmont Hotel.

Once back home I started my search for the perfect antidote. Considering how many toxic chemicals there are, it was not easy, but I wanted to give the best advice to BK that I have ever given to any client. You always save the best for your family and friends.

After all, isn't that what friends are for?

In my quest for an antidote I decided to seek reliable advice. I am fortunate to know some of the best and brightest health professionals, so I reached out to my network. I asked my friend Dr. Zack Mansdorf, past president of the American Industrial Hygiene Association and recently retired Corporate EHS Director for the L'Oreal Corporation, for ideas. As hoped, Dr. Mansdorf came back with a golden nugget of information. This is what I learned from Zack (with thanks to Wikipedia):

> The universal antidote used to be a mixture that contained activated charcoal, magnesium oxide and tannic acid. All three components were thought to neutralize the actions of various poisons. It was prepared by mixing two parts of activated charcoal, and one part tannic acid, and one part magnesium oxide. This cocktail was administered to patients who consumed unspecified poisons. However, this mixture is considered marginally effective and is no longer used.

Current thinking, according to Dr. Mansdorf, is that just plain old activated charcoal in a ratio of 8:1 is sufficient. This means 8 parts of activated carbon for every 1 part of poison. Whoever needs to take this antidote should mix the carbon in water to swallow the paste-like cocktail.

My friend BK Nor was in luck. If he ever suspects that he has been poisoned, he can take an organic vapor cartridge, crack it open, and just swallow the granulated carbon. BK is in the PPE business, so he has easy access to plenty of carbon cartridges.

I no longer worry about my friend BK.

CHAPTER 8

DANCING DEBBIE

People's Republic of China

In 1989, an organization called People to People International (www.ptpi.org) asked me to lead a delegation of professional industrial hygienists to the People's Republic of China. The delegation included Americans, Canadians, and a lone European. We were scheduled to visit several big cities in China and make technical presentations to our professional counterparts during the pre-planned events. Also planned were numerous cultural events, sightseeing trips, banquets, and social networking opportunities. Being seven thousand miles away from home for fifteen days, we would have plenty of time to make new friends and experience Chinese life. As the delegation leader, my expenses were borne by the sponsor.

To say this sounded good would be an understatement. I thought I had died and gone to heaven. I pinched myself and then pinched myself again a second time, but even harder. When I realized that I had hurt myself and that I had done few things worthy enough to merit heaven, I came back down to earth. Still I could not help but be excited about this once-in-a-lifetime trip.

Please be a little less happy

Our first stop was Beijing. We stayed at the venerable Beijing Hotel, only a hop, skip, and jump from Tiananmen Square, where only one week after our departure the historic and tragic events in the soon-to-be infamous square unfolded.

After the long flight from the USA, a cold Qīngdǎo (also called Tsing Tao) beer served in the hotel lobby provided a welcome respite. About ten of us were sitting in the ornate lobby bar speculating about the coming two weeks in China. After each round of Tsing Tao beer, the ethanol-fuelled conversation turned louder and more boisterous. Soon we became a nuisance and no doubt were disturbing other patrons.

The lobby bar manager, a bespectacled middle-aged man, was troubled. He started hovering around us nervously. Something was on his mind. When he could wait no longer, he walked up to us and in barely audible tones, said, "Ladies and Gentlemen, can I ask you please to be a little less happy?" To which we understood as saying, "Please do not be so loud and obnoxious. You are disturbing others." We lowered the volume and promised to be less happy. That made the lobby manager more happy. This valuable introductory lesson in Chinese etiquette served us well for the remaining days of our China trip.

Shanghai disco

Debbie was the youngest delegate in the group but certainly not the quietest. She soon became the group favorite and a sought-out companion at dinner tables and the numerous sightseeing trips. Debbie also claimed a seat on every unsanctioned side trip or activity labeled: "Not Recommended for Foreigners."

Debbie was about thirty years old when she went on the People to People International China trip. This was one of her first international trips because she was a relatively new employee at XYZ Inc., a small, privately-held corporation. XYZ was founded in the early 1960s by an entrepreneurial scientist who saw the global need for air sampling equipment for the fast-growing industrial hygiene profession. Despite its humble beginnings in an old barbershop, XYZ grew quickly to be a global market leader known to every industrial and occupational hygienist in the world. Debbie's contribution to XYZ's vast global presence was well recognized by her colleagues, competitors, and peers.

After Beijing, our next stop was Shanghai. Shanghai was a different city in 1989. None of the glamorous skyscrapers of today were there—no World Financial Center, magnificent Jin Mao Tower, Oriental Pearl Tower, or the Tomorrow Square housing the sparkling JW Marriott Hotel. Shanghai had only a handful of international hotels, ours being one of the few. Unlike today, there was not much night life in the city.

Rumors spread that a new disco had just opened near our hotel. It was the fanciest club in town and popular among the Shanghai elite. Debbie sprang into action. She found out all the details. She arranged the transportation, and soon a dozen of the disco ducks were on their way to the Chinese disco.

The place was chic by the Shanghai standards of the time. The music was a mix of the seventies and eighties and included teen, pop, disco, jazz, swing, hip hop, and soul music—pretty much everything I didn't listen to back home. Michael Jackson's songs were popular. Local couples, dressed to the hilt, first seemed hesitant to step on the dance floor, but after they had a drink or three, they started to migrate to the dance floor. As the beat picked up, the dance floor really started hopping. Debbie jumped onto the dance floor and motioned vigorously for the rest of us to join her. Several of us obliged. The Chinese already on the floor evidently approved, judging by their smiles. It was great. Soon a strange thing happened. One by one, our new Chinese dance friends started clearing the floor. Within a couple of minutes, only a few of us foreigners were gyrating.

Debbie seemed worried. She inched closer to me and whispered in my ear: "Maybe we should leave. I don't think they want us here. Look, they all cleared the floor as soon as we stepped in. It is obvious we are not wanted here."

"That is awful," I said. "We were told that people in Shanghai are friendly. They like to meet foreigners and practice their English language skills. I do not understand this. This is bizarre and insulting."

As I was finishing my sentence, I noticed an amazing scene unfolding. I moved closer and told Debbie, "Hey, look, nobody has really left. They are all still here, almost in a circle. Just watch, everyone is looking at your feet and your movements. Do you realize that they are imitating your every movement? They are learning from the young master."

"Oh my God! This is amazing. I've never seen anything like this," Debbie screamed above the pulsing music. A big smile radiated from her vibrant face. With an energetic wave of her hand covering a 270 degree arc, she motioned to the crowd to join her. I looked at the crowd, gestured to Debbie, and yelled, "Introducing Dancing Debbie!" The crowd reciprocated by enthusiastically chanting, "Dancing Debbie, Dancing Debbie . . ." They clapped and hopped onto the dance floor, and we danced till the wee hours of the morning.

Debbie reached out and hugged the young spectacled Chinese man who had led the chant. His pale complexion turned blood red in a matter of seconds. The bear hug from a beautiful, aggressive American woman was not in his horoscope. The gesture startled the young man beyond recovery.

The nickname "Dancing Debbie" has since been heard around the world, including in Australia, Singapore, Sao Paulo, and Israel.

Bourbon Street, Brazil

In 2011 the annual meeting of the Brazilian Occupational Hygiene Society—the Associação Brasileira de Higienistas Ocupacionai (ABHO)—was in Sao Paulo. This was my first time attending this sort of meeting in Brazil where industrial hygiene was becoming more popular. Debbie's company, XYZ, was well represented. Besides XYZ's local Sao Paulo crew, Dancing Debbie and her senior partner from the USA headquarters were also present. Knowing that Debbie would be in town, a search was begun for the best salsa dance floor in the sprawling city. It was an easy task. XYZ's local leader, Reinaldo, knew all the good places in town. He invited several people from my company to join them at the well-known Sao Paulo dance club called "Bourbon Street."

People usually do not refuse Debbie's party invitations. Ask anyone who has heard of Debbie's party at the annual conference of the American Industrial Hygiene Association. The invitation to the annual XYZ party does not come easy. First of all, it is for "select foreign delegates." As an adopted Hawaiian, I do not typically qualify as a foreign delegate at a conference organized by a US company. I must have done some good deeds to receive the coveted invitations from XYZ for the past fifteen years. I have thoroughly enjoyed this status even though I don't recall earning it. Whether I earned it or scammed it, I relish my standing invitations. This has allowed me to take a guest of my liking every year to Debbie's party.

The caliber of the people on the Sao Paulo dance floor was high. They seemed like individuals handpicked to dance at this exclusive club. I love Latin dancing, but there was no way I was going to get on that floor no matter how many *caipirinhas* I drank. (Caipirinha is Brazil's national cocktail, made with cachaça [sugar cane rum], sugar, and lime.)

I went to the Bourbon Street club with the people from my company. The XYZ gang arrived around the same time. Debbie showed up a few minutes later with a tall, flamboyantly dressed Latin dude. His name was Humberto. Humberto looked partly like an aging Casanova and partly like a fun-loving Chicago gangster. He stood out in the crowd and made an impressive entry. Debbie introduced him to all of us.

"This is Humberto, my dance instructor. Humberto is a native of

Bolivia living in Brazil. He is the best, and I am so happy he is my teacher," said Debbie. It was clear to me that the Bolivian dude had cast a spell on Debbie. She said Humberto was the salsa king (El Rey), and dancing with El Rey at Sao Paulo's famous night spot was living her dream. Many eyes were focused on Humberto and Debbie. I cannot be sure which one of the two was the recipient of the most glares.

One of the Humberto–Dancing Debbie moves that attracted a lot of attention was when they danced back-to-back. The move is called "butt kiss." None of our Portuguese colleagues confirmed the nomenclature but that does not matter. It was a good description of the activity. Debbie encouraged me to try the step. I declined. This was way too advanced for me. Additionally, I was not sure if she wanted me to learn the "butt kiss" from Humberto. Plus, all of my friends had cameras.

After Debbie had introduced all of us to the Salsa El Ray, she turned to a young, somewhat shy girl, pulled her close, directed her our way, and said, "Hi everybody. I wanted to share the joy of dancing with my friends. So I have hired this woman for Jas for the evening."

I screamed, cupped my face between my two hands, and shouted: "I already have a woman. I do not need more than one. For one thing, my religion would not allow such luxuries. Plus, I am happy as is."

Everyone laughed. "No, silly. Not that. I meant to say I hired a dance instructor for you. You have always told me you wanted to learn some Latin dances, and so now is your chance. Reina is the best and the most authentic. She is Humberto's sister."

Reina did not look like Humberto's sister. I am certain she was the king's girlfriend, but that did not matter.

Debbie continued, "Reina attended a university in the USA for three years and now she is in Sao Paulo teaching Latin dance to her American students, including businessmen, diplomats, conventioneers, and others. She is a business partner with Humberto. Reina will be your personal dance instructor."

"My personal dance instructor? I have my own dance instructor in Sao Paulo, Brazil. What did I do to earn this?"

"You gave me the nickname 'Dancing Debbie,' and that has inspired me to pursue my passion even more vigorously," Debbie assured me.

As soon as the music started again, Reina grabbed my left hand and steered me to one corner of the floor to avoid bumping into other people. In the corner, not too many people would notice an awkward novice stumbling among the best dancers in town. Reina was very patient. She did not frown every time I pulled in the opposite direction from the one she intended. It was a tug-of-war. As I danced, I was reminded of a mother holding a toddler's hand, wanting the child to go in one direction while the child would pull in a different direction. I could not help myself. I felt so many distractions. So many beautiful people were dancing around us. Nevertheless, Reina was still very patient with me.

We were constantly counting. To keep the timing with some of the intricate Latin moves, you need to count your steps and pronounce the numbers. I had to repeat one, two, three, one, two, three so many times that I thought I was back in kindergarten. When the music stopped, I headed toward the exit. Reina stopped me and asked me to wait for the next dance. After two more dances, Reina said, "You want to sit down? You seem tired. Let me know when you are ready and I will take you out on the floor again."

I was tired of counting and stepping on her toes. Watching other dancers on the floor was a lot more fun. I realized that I was not interested in many more lessons. "I am not tired. I just want to watch how the experienced people do the steps you were trying to teach me. Let us just watch for a while."

I did not go back on the floor. Debbie came to me and said, "What is the problem? Reina thinks you don't like her."

"It is not a matter of liking or not liking. She is lovely and talented. I just feel very weird taking lessons next to such accomplished dancers. A high school classroom is a more appropriate locale for me to practice," I said.

"Okay, but do one more dance with Reina so she will not feel that she failed," insisted Debbie.

I danced one more time and actually noticed some improvement. Both of us felt some sense of accomplishment. I thanked Reina and told her she was a lovely teacher.

Debbie danced with Humberto the whole evening. She wanted her money's worth and the money's worth she had spent on Reina to teach a few steps to her friend Jas. I thanked Debbie. I do not have many friends who hire friends for me for an evening. This is probably good, I suppose.

Cuba Libre, Singapore

The American Industrial Hygiene Association (AIHA) had its maiden Asia Pacific Industrial Hygiene Conference in 2012 in Singapore. The AIHA felt that the Asia Pac region was ripe for industrial hygiene and that Singapore was a good location for a high quality professional IH event. The 2011 conference was very successful and seems to have become a regular event of the association.

Debbie and her company were well represented at the Asia Pac conference. As soon as Debbie marched into town, feelers were sent out to locate Latin dance clubs in Singapore. This was a tall order, finding an authentic

Latin dance club in Singapore, but as they say, where there is a will, there is a way. One of the managers in our office recommended a place called "Cuba Libre" and rounded up a dozen people to dance to the Latin beat in Singapore. At first I was skeptical. Dancing Samba and Meringa to an Indian Tamil (Tamil Nadu State in India) band didn't seem right. The band was good. It was not the Bourbon Street in Sao Paulo, but they did a credible job. Debbie danced nonstop with almost everyone that night. Before long she was sweating profusely. When she finally left the floor, we heard clapping and muted shouts of "bravo," Singapore style.

Another one bites the dust.

How much is that dress in the window?
At the AIHA conference in Indianapolis in May 2012, Debbie told me that dancing was no longer just a hobby for her to relax from all her industrial hygiene duties of marketing air sampling instruments. She was into dancing in a big way. All this did not come cheaply. She had to cough up five big ones ($5,000) for a dress so she could dance with the big boys.

She turned to my friend Brian and said, "Would you fellows care to see my new Catwoman dancing dress?"

"You will change into your dress while you are on booth duty here in the exhibit hall?" we both exclaimed in unison. We were aghast!

"Don't you wish," she laughed. "No such luck, guys. I am at work, but let us move to the corner so as not to interfere with the other sales people serving the customers. I will show you the dress I was wearing at the last Catwoman dance I performed."

We watched a video of the Catwoman posing in her dance outfit on her notebook computer. The Catwoman looked awesome in her $5,000 dress that had very little fabric but lots of sparkles and laces.

"Latin dance dresses are like Latin lovers—hot and sexy. These dresses tend to be more like bathing suits with a little extra material for style. Every dance dress has two critical components—slits and sparkles," she explained.

The Catwoman's dress was no exception. The aqua blue sensation had peek-a-boo windows in all the right places, and crystals sparkled from top to bottom. What could I say? "That is nice."

"Where is the fabric? I only see some ribbons, laces, and sparkles. How do you cover the skin most susceptible to toxic exposure that requires some form of PPE (personal protective equipment), for skin exposure in this case?"

"Remove your sunglasses so you can see, Jas. You are inside a building. You have nothing to hide here." (I have a lot to hide, but I did not want to bare my weaknesses in front of so many people.)

I took off my sunglasses. "Okay, I see some fabric, but in my way of thinking it will still leave areas vulnerable to chemical and physical assault. The costume does not provide sufficient coverage." Without the benefit of a tape measure, I quickly estimated Debbie's total surface area that needed to be protected to be 0.66 meters.

Debbie said, "When worn, it stretches and covers everything worth covering. You will be convinced when I show you the video where I am wearing it in the Catwoman dance."

"Okay, here it is." Debbie adjusted the computer screen and explained the Catwoman's moves. She was right. The dress did cover the skin most prone to hazardous exposure. Debbie looked like a seasoned professional dancer. For the moment Brian and I both forgot that she was a CIH attending an exhibit demonstrating the devices designed to measure toxic chemicals and poisons in the air.

The video of her as the Catwoman seemed to be from another world. It was great chemistry. It was chemical and physical magic in the hands of a Catwoman who could resurrect the lover she had just killed with her touch. Catwoman is passionate yet she also has a temper. Watch out!

Dance till you drop

A few years back, Debbie suffered from a brain hemorrhage due to an aneurysm. She almost died. Her situation looked bleak, but she was not ready to go. Her best days were ahead. Dancing Debbie wanted to keep on dancing.

Amazingly, Debbie danced not long after two brain surgeries. She landed on her feet. After a few weeks in the hospital and at home, she was ready to roll, albeit slowly at first. This was indeed a life-changing event that propelled her to reevaluate priorities. Debbie recommitted her life to the Good Lord and asked for His blessings, especially with her dancing. It was clear that the Lord did not want her to go quite yet. Life is precious. It must be lived. Debbie must dance.

Soon afterward, as luck would have it, a ballroom dancing studio opened within a couple miles of Debbie's home in Texas. Better equipped with high energy and a decent credit limit on her charge card, Debbie plunged into Latin dancing deeper than ever under the tutelage of a handsome man half her age. The Good Lord must have intended it to be that way. During one recent lesson, she reverted back to her days as a cheerleader and fell to the floor in a perfect split. Getting up was not so perfect, as she had torn a hamstring. Six months in chiropractic care were required to keep her dance career moving forward.

Recently she won two gold medals in dance contests. More are in the works. I am happy but worried. I am glad to see my friend achieve what she dreamed about all her life—dancing. I am afraid that the industrial hygiene profession someday will lose one of our most enthusiastic ambassadors. What would XYZ be without Debbie? Would the annual IH conference be the same without her if she could not make the conference because of a dance conflict?

So this is an alert to future host cities for AIHA conferences. I know that Debbie will be there. The Catwoman will be ready to tango, rumba, salsa, swing, and merengue. She has the dress. Please make sure that you have Latin dance clubs. If you do, please get in touch with Debbie. Do not forget to cc me.

You are invited!

CHAPTER 9

LOST IN TRANSLATION – PART 1

Dr. Wu may be the nicest man in China. He is pleasant, polite, and humble. In fact, he is so polite and so humble that it obligates his friends and associates also to act polite and reserved and show restraint when the situation calls for one to be irreverent. In Dr. Wu's presence, to act otherwise would appear vulgar because he is so polite and proper. If I could share a visual picture of Dr. Wu with you, you would think I had just conjured up some mythical character and the picture that I was showing was actually the revered Dalai Lama. Dr. Wu has an uncanny resemblance to the spiritual guru. Just looking at Dr. Wu makes you want to bow to him. This is difficult for me because Dr. Wu and I are close friends. I can tell Dr. Wu some very weird things. I can also tell him jokes that you would not dare tell in mixed company. Perhaps because we are friends, he does not seem to object to my bawdy, tasteless jokes or to my sometimes X-rated vocabulary. He just smiles and dismisses it all as if ordained by the Supreme One to act in this fashion.

Fortunately and of necessity, most of our discourse is about health and safety issues, not gutter talk or scatological musings.

Do you want some company?
The 2004 annual conference of the American Industrial Hygiene Association (AIHA) was in Atlanta, Georgia. Dr. Wu and I stayed in downtown Atlanta near the convention center. For me, over the years, the AIHA's annual

conference has become a rendezvous point, a kind of family reunion and a networking and business development opportunity rather than a pure technical gathering. To a certain extent this is unfortunate because I miss out on much of the technical discourse. Nevertheless, I love the event. Come rain, come shine, and sometimes snow and hail, I have not missed the event for the last thirty-nine years. Family reunions, birthday celebrations, funerals, and engagement parties can wait or be moved around, but not the annual AIHA conference.

Tuesday night at the Atlanta conference, Dr. Wu and I met for dinner at one of the fine downtown eateries. To make it even more enjoyable, Dr. Wu insisted on paying the bill although he had no reason to pick up the tab for our party of five. After some mild arm twisting, I let it go and allowed Dr. Wu to pay the bill. He was a generous person and I was not willing to engage in an aggressive power play with someone who looked like the reincarnation of the Dalai Lama.

Dr. Wu and I said good night to our dinner partners and started walking toward the Holiday Inn where we were staying. It was past eleven p.m., and around that time Atlanta's streets take on a different character, like many central city streets in the big American cities. Strange and exotic nocturnal creatures come out after eleven p.m. Normally I would avoid such excitement, but our hotel was very close to the restaurant we had just left. Taking a cab would look ridiculous.

Around the corner from the restaurant, Dr. Wu slowed down a little to retie his shoe lace when a shapely and provocatively-dressed woman caught up to him and whispered: "Would you like some company tonight?"

Dr. Wu, who had not yet finished the lace-tying task, straightened up in a hurry to answer the query. In all this confusion, the only word he understood from the young lady's offer was the word "company." He gathered himself and responded:

"Company? Company?" He then went on to answer his own question, as some people often do. "Yes, of course. ABC Environmental in Michigan."

The young lady looked at both of us like we were a strange species just dropped by an alien space ship into the middle of downtown Atlanta.

"Do you want some company?" she repeated to Dr. Wu, only slower and louder this time.

Wu was ready with his rehearsed answer.

"Yes. ABC Environmental from Michigan."

"What the fudge?" The young lady uttered a familiar profanity (which I have altered for this story), convinced more than ever that we were not of this planet. She must have thought we were not from planet Earth because on Earth everybody understands English, including the kind they speak in downtown Atlanta, Georgia, USA after eleven p.m. She moved away from us as fast as she could, still uttering obscenities.

Dr. Wu was devastated. He turned to me and said: "Jas, why does the lady dislike our company so much? The minute I said 'ABC Environmental,' she got upset. I did not mean to upset or insult her." Wu was visibly disturbed and apologetic.

"Don't worry about it. It may have nothing to do with our company but, on the other hand, it is possible that some guy from our company stiffed her. Such things happen during big conventions," I tried to reassure Dr. Wu.

"What do you mean by 'someone from our company may have stiffed her'?" Wu was inquisitive as usual in his perpetual pursuit of knowledge.

"What this means is perhaps someone used the service and refused to pay the invoice." I continued, "You know, Wu, it is like you give your body and soul to serve a client on a difficult assignment and then some deadbeat refuses to pay the bill. Take our own situation. We do good work and some clients refuse to pay us."

"Oh, I can understand that. Very bad. It is very bad. Not good for a company's reputation."

"Damn right!" I said but later regretted using the word "damn" in front of the Holy One.

As usual, we were both in complete agreement.

I miss you too

Nineteen ninety-seven was a transitional year in my career. After twenty years with the same company, I was seeking change. Eager to practice my trade in Asia, I accepted a job from a big US insurance company to run a rag-tag risk management group in Kuala Lumpur, Malaysia. The new assignment was every bit as exciting and different and chaotic as I could have imagined. I was ill informed and ill prepared to take on this assignment. The company in

Malaysia that I took over had a lack of organization and also lacked the means to adequately traverse the cultural, religious, and legal formalities required to do business in this ethnic mosaic. Despite this, the assignment turned out to be a delightful adventure for me: friendly people, good food, warm climate (sometimes too warm), and a surprisingly high level of living standards. The group I inherited included Muslims, Christians, Buddhists, and a Sikh. Most were Malaysians, but one was a Scotsman and one was a blonde Swede with the name Ali. All in all, a nice bunch.

My immediate need was to hire a few more people, the engineer and science types. My local second-in-command advised me on how to find technical people (place an advertisement in the Saturday newspaper), what to ask for in the advertisement, and what to expect from the job applicants. Soon I was getting loads of resumes. The information contained in the resumes was quite different than what I was used to back home in Orange County, California. Each resume came replete with personal details and a photo. Information supplied included:

Name:

Sex: (This always amused me. I thought the more suitable word would have been "gender," but then perhaps the word sex has more common appeal.)

Date of birth:

Age (including months):

Religion:

Race:

Marital status (married, single, divorced, on hold):

Height in meters and centimeters, sometimes to the decimal point:

Number of children: (In India, the job application would ask for "number of issues" because for official purposes, your progeny are referred to as "issues." This term has intrigued me for years but I understand it now. If you are poor, the children, no matter how much you love them, still mean more mouths to feed so it can be a significant issue in your life.)

It was hard for me to digest and properly apply all this information in making informed decisions. For example, how to deal with a candidate's height? I had no guidelines or thresholds to consider. At what point do you reject a candidate on the basis of height? Was height a factor in order to be a good consultant? Do employers reject applicants because someone is a couple of centimeters too short or maybe too tall?

Taking into account the anthropometry of the Southeast Asian cohort, I was thinking that back home if I sought such data from applicants, I could be hauled in front of some kind of equal opportunity commission or department. Actually, earlier in my career I was summoned by the Human Rights Commission in Syracuse, New York, for discriminatory hiring when I tried to select a junior stack testing technician. It was a frivolous claim. I came out smelling like a rose from that hearing and was absolved of all crimes.

My advertisement in the Kuala Lumpur newspaper brought in interesting applicants. Among them was a young Malaysian lady of Chinese extraction. Her qualifications seemed to match our requirements better than most of the other applicants.

Bachelor's degree in chemistry

Four years of experience

Bi- or trilingual (if you consider Hokkien a separate and unique Chinese language)

Single

Five feet tall (she gave her height in meters, but I had to convert it into feet and inches. I can never visualize an attractive woman in meters.)

Her name was Marina Yu. The obligatory photo attached to her application showed a pleasant, composed, and confident young woman.

"This is our candidate," I confided to my second-in-command, a woman of discriminating tastes and the kind not easily impressed.

"So soon, Jas? That is it? Isn't this a bit fast?" She was questioning my objectivity without saying so. She continued, "Aren't you even going to call her for an interview?"

"Oh yes, of course." That was implied. I quickly tried to reestablish my objectivity. I don't think it worked.

The following week Miss Yu came to the office for an interview. She proved to be as good a candidate as I had imagined. However, I did not offer her the job on the spot. I wanted to be fair, and most of all I was not going to fill the place with candidates whose bio data I admired. A couple of other candidates with good academic backgrounds and good bio data had surfaced in the meantime, and I wanted to meet them before offering anyone the job. I thanked Miss Yu and promised that I would call her in two weeks.

Two weeks later, I called as promised.

"Miss Yu," the voice at the other end declared firmly.

I had heard phones answered like that before. One of my business associates, when called, will only answer "Roger!" and nothing else, leaving you gasping for more or wondering how he knew my name was "Roger" (which it isn't). Another one named Jim Pierce will pick up his office landline and say, "James here." I always wondered why he said "here"? I knew he was there. I had called his desk. Why would he need to specify his GPS coordinates and explain his whereabouts? With caller ID on cell phones these days, perhaps such information may be relevant, although I still don't need to know if the person I am talking to is in their office, at the shopping mall, or, more likely, at Starbucks.

Anyway, when the lady on the other end of my call said, "Miss Yu," I got confused. It was still early in the morning and I had not slept well the night before. Moreover, Miss Yu sounded exactly like my wife back in the USA. Without thinking, I said:

"Yes, dear, I miss you too."

No one slammed the phone down, but there was a long silence and Marina finally said:

"Who is this?" The anger in her voice was evident. Apparently she had been getting too many crank calls lately. She decided that it was time to confront the offender.

"Miss Yu, this is Jas Singh. I am following up on your visit a couple of weeks ago to our offices and, before I say anything else, please allow me to apologize for the odd comment. Actually it was not really odd considering the fact that I thought you were my wife." I stumbled through the awkward statement. It was not perfect, but it worked.

"Oh, thank you, Jas. I am so glad I didn't hang up the phone because I was sure you were a prank caller. There have been many lately."

I thought of blurting out: "And how do you know, Miss Yu, that I am not a prankster?" but something restrained me. Such humor is not appropriate with a potential junior hire you have met only for an hour and under a very formal and somewhat tense environment.

"Miss Yu, we have finished our interviews and have decided to offer you the position. Congratulations. I will send you the formal offer. You will be an asset to our company, and we look forward to you joining the team."

Marina was delighted. I felt like she was ready to give me a warm hug (maybe my imagination, a too-common phenomenon). Regardless, her delight at the offer was genuine and obvious.

Miss Yu turned out to be quite an asset. We went on dozens of field projects in several Asian capitals. The best trips were to China because she spoke the Mandarin dialect, a heaven-sent gift when travelling in the Chinese hinterlands.

I often run into Miss Yu when in Malaysia or at health and safety conferences in Asia. She is an accomplished health and safety consultant.

The psychotherapist

Dr. Rajiv Gupta (not his real name), is a well-known psychotherapist in New Delhi. In a country where basic medical needs are hard to come by, psychotherapy may not seem like a flourishing area of practice. Rapid growth of the Indian economy, however, has spawned a crop of nouveau riche people who do not know how to handle their newfound fortunes and need psychiatric help to deal with the stress.

Stories abound about people doing some strange things to show off their newfound wealth. A widely reported story tells of a farmer in the industrialized suburbs of Delhi where land prices skyrocketed because of all the multinational companies moving into the area. The farmer, who owned many acres of land, had so much cash from his property sales that he wanted everyone to know of his riches. At his son's wedding, the farmer chartered a helicopter to make a few rounds over their village so everyone could see him and his son riding inside the flying machine. After the planned few rounds, he was not satisfied. Peering down, he noted that some villagers probably missed their flyby. He ordered the pilot to go around a few more times, agreeing instantly to increase his contract by a few thousand US dollars.

Such anxiety causes much stress and the need to seek psychiatric help.

Dr. Gupta's practice took off along with the Indian economy. He rented new plush offices and moved out of his dingy quarters in old Delhi. At the same time he found new love and married a comely educated Delhi socialite. The two decided to go on a honeymoon to Europe—I am not sure where, but the rumor was Switzerland.

To advertise his new location upon his return, Dr. Gupta ordered a big, flashy, neon-lit sign that would read: "Dr. Rajiv Gupta, MBBS, PhD, PSYCHOTHERAPIST."

Before leaving for Switzerland he checked the billboard design with his painter and agreed on the colors, height of the letters, and the exact position where the billboard would hang above his entrance door. He told the painter that the sign must be in place and lit when he came back from his honeymoon. The painter was the best calligrapher in town. Spelling, however, was not his strong suit. To make the sign more evenly spaced, he decided to split the long, awkward "psychotherapist" word into three parts and hung the sign as instructed.

Monday morning on the appointed day, the good doctor returned from his honeymoon and went to the office early. He had several important appointments waiting for him. As soon as the doctor approached the office door, he froze. He looked at the sign and relooked and could not believe his eyes. He started yelling for his assistant, who had just arrived a couple of minutes before. "Take the stupid sign down. Right now! I mean at this very minute." The good doctor was screaming and was personally in need of psychiatric help. The obliging assistant ran to grab a ladder and swiftly removed the bulky neon sign.

The sign read:

Dr. Rajiv Gupta, MBBS, PhD.
PSYCHO THE RAPIST

CHAPTER 10

LOST IN TRANSLATION – PART 2

I met William six years ago at the Shanghai JW Marriott Hotel, a magnificent property in the heart of the city. The JW Marriott was an unlikely place to recruit a mid-level occupational hygienist. This hotel was also a notch above my budget, but I thought why not? My client was staying there and he had booked the hotel for me under their corporate account.

Our China manager had asked me to help him interview William and another candidate for a position in our newly created industrial hygiene division. William had been working at a local government occupational hygiene agency called the "CDC," the same acronym we use for the well-known US agency called the Center for Disease Control in Atlanta, Georgia.

I was immediately impressed by William's grasp of industrial hygiene, not a well- established discipline in China at the time. William was different. He rattled off every chemical exposure limit established by the US Occupational Safety and Health Administration (OSHA) I had mentioned. He knew who the leaders were in the industrial hygiene profession. Then he threw a zinger at me when he said: "Dr. Singh, I know of you!"

"Not possible," I said. "I have not done much work in China and I have not been publishing anything that you would have come across." I looked at William to determine if he was bluffing. One reason I thought he might be bluffing was that he could not hide his sly smile and the twinkle in his eyes. Now I know that he wasn't trying to be sly or clever. That was his natural look.

"You are famous in China, Dr. Singh. You may not realize this."

I instantly liked the guy. He had found the key to my ego. He had discovered my weakness for flattery with lightning speed. I said to myself, *If this guy can read me so quickly, imagine how successful he will be with clients and prospects.*

For me the interview was over shortly after it started. He was our man. I did not need to know what courses he had taken or how many poisonous gasses he had tested in his short career. I was going to recommend to the China manager that we grab him. The interview, however, lasted another hour. The China manager had many more questions of William. I could not blame him. After all, he was the one who would have to pay for William's dim sum.

The mentoring process

As expected, I was given the assignment of being William's mentor. It was a pleasure. You couldn't ask for a better assignment. William is intelligent (I say "is" because he is still with us and I don't think his intelligence has diminished with time), quick, and humble. Anything I told him, he gobbled up and came back for more, so much so that after a while, I started feeling insecure that pretty soon the guy would know everything I knew and I would have little left to teach him. This concern was so paramount that a few times I found myself contemplating sandbagging (withholding information from) him. But that would surely have been unethical.

William kept reminding me: "Jas, you are the best mentor anyone can have. You have such a reservoir of knowledge and real-life experiences." I could never figure out if he really believed all this, but it made me feel good. Still I could not get rid of the feeling that I might run out of ammunition someday where William would not need me anymore.

Suddenly one day I realized that my insecurity was ill founded. I had another treasure trove in my arsenal that William might need. This was about American customs, American informal greetings, jokes, slang, and curses that I could teach him. William had an insatiable appetite for such knowledge in his desire to fit in anytime he was in the company of Americans at professional conferences, especially the annual conference of the American Industrial Hygiene Association.

This was good news to me. I could mentor him, although I soon realized that in a way this arrangement was rather weird—an Indian man teaching American slang to a Chinese colleague to impress American ladies (this is not my specialty, but I think I am a little better at it than William was because I had more practice). With that we went to work with this new mission in mind.

It is a cake

Slang and curses notwithstanding, what I really wanted to teach William was American business communications, especially the importance of writing technical reports that were short and to the point. I had invested much time in my early years to learn this skill and I wanted to share this with my friend. I told William about the course I took that changed my life. It was a four-day course taught by a famous professor from the New York University School of Business. The man had a unique but effective style. In just four days my writing style changed forever. He made me literally swear not to use certain words and expressions. One of the things he drilled in my head was the value of brevity.

"Get rid of all the chaff. It adds no value." He told me my reports were too wordy, had too much chaff, and my style put them neither in the scientific realm nor in the literary category. He also told me to write in an active, not a passive voice. "Get out of that nasty habit of writing in the passive voice," he would shout. "The people in your profession (engineers, scientists, and consultants in particular) are inflicted with this disease, and do you know why this happens? This is a reflection of people's insecurity, lack of confidence, and lack of conviction in their results and recommendations." His theatrics worked on me. I swore not to use passive language, to cut the chaff, and to be brief. I am still working at it. It is not easy.

I decided to share this advice with William. He was very excited. He wanted to learn how to write brief and precise technical reports in the

American style. Like everything else, he was gobbling up all this wisdom like he gobbled technical details.

The proof came soon after. A client called me from Wisconsin. People at their plant in China were experiencing dermatitis problems. It was a complex investigation and required a medical professional in addition to an industrial hygiene expert.

I sent an email message to William, asking him to give me his honest opinion if we were qualified and equipped to do this job.

As usual, the answer came within four hours. "Jas, it is a cake." Just those words. Nothing more. Why say "piece of cake" when just "cake" will do?

Lesson well learned.

The mentoring process was working.

Legs are nice

Despite my efforts to "Americanize" William by teaching him colloquial and not always polite expressions, William remained a reserved man yet still insisted all the while that he wanted to know the rough and ugly side of life, too. So I kept up my efforts to tutor him but in a measured way. One delicate issue was how to show appreciation for a beautiful lady when she walked by without uttering uncivilized remarks, catcalls, or rude gestures.

The opportunity soon presented itself. We were standing in a public area when a fashionable Shanghai coed suddenly appeared from our right. Dressed in a provocative, eye-catching dress, she personified the ultramodern Shanghai woman of today. Heads turned from every direction as she walked past, some twisting at an angle that my ergonomist mentor, Dan McLeod, would describe as a "hazardous posture." I was hoping for some reaction from William. Nothing. Absolutely nothing! His head did not turn even the ergonomically safe 15 degrees. The only motion that betrayed his curiosity was a 45-degree diversion in his right cornea indicating that he was not totally oblivious of the stunningly beautiful lady who just walked past us. But that was the extent of his reaction—no words, no facial expression.

I was dismayed. I had failed as a teacher of social behavior and American etiquette. All my efforts to Americanize William were in vain.

I could no longer contain my frustration. I looked at William and said: "Nice legs."
"Yes, legs are nice," William muttered without ever lifting his eyes off the road and looking at me or the pretty lady.

Legs are nice! What a profound statement, I thought. This could only come from a Chinese immersed in the Confucius teachings. *Faces are nice too,* I thought to myself.

I turned to William, determined to draw something more out of him and said: "Of course, William, legs are nice. If they weren't nice, the Good Lord would not have given us legs, and can you imagine if the Good Lord did not give us legs? We would have no legs to stand on."

This brought William to full attention. He looked at me with a mixture of surprise and dismay and said: "What in the world are you talking about, Jas?"

This was encouraging. I knew he came very close to saying, "What the 'hell' are you talking about?" but he did not. Saying that would have been rude in William's world, especially when addressing one's mentor.

The brief encounter and the discussion following had a long-lasting effect. After the episode, every time a young lady appeared wearing a short dress and bare legs (plenty in Shanghai these days), William would turn his head ever so slightly and whisper, "Nice legs," even if the legs were crooked and spindly and not so nice. But who cared? The lesson was thoroughly absorbed.

Spell-check is not reliable

William and I spent countless hours on report reviews, a crucial part of our jobs. Our diverse workforce, with varying technical and ethnic backgrounds, was always a challenge. William had several very eager and bright young men and women working for him. For reasons totally comprehensible, their mastery of the English language did not match their technical prowess, especially the American dialect of the English language. Being a North American company, all of our manuals, procedures, and guidelines were in the American language. This always confounded Asians, who had years of English influence and were now being bombarded with American culture, including the language that Americans and Canadians spoke. The expression "eh," a common utterance by our Canadian colleagues, further added to this confusion.

For example, an industrial hygiene report on a project authored by a new employee we shall call Ryan (probably the most popular Western name in China, second only to William) cleared every quality assurance step before it got to me. It was a simple project. Air samples were to be taken in the vicinity of a worker's "breathing zone" and as close to his nose as possible to simulate what the person was breathing when working with certain toxic materials. Ryan faithfully recorded the process, the methods, and the terminology, following the procedures as prescribed by the book. I was impressed with the report and did not find any faults. William had already viewed it with his built-in magnified glasses. Suddenly, I came to a screeching halt. One word in Ryan's description of sampling methods caught my eye. I read it and reread

111

it. Ryan misspelled "breathing zone," and it had gone undetected through the latest version of the spell-check software. His intent was to declare that as proscribed under the protocol, "samples were collected in the worker's 'breathing zone,'" but unfortunately, his sentence read: "The samples were collected in the worker's 'breeding zone.'"

"Oh no!" I screamed.

Luckily, the report had not been issued because it had not yet been through the final QA check. I called William and asked him to look at the paragraph describing the sample collection method. He looked at it and asked: "You see a problem?"

"Yes, I do. Look at the spelling of the 'breathing zone.'" After fifteen seconds I heard an echo of my "Oh, no." This was followed by an awkward silence, then simultaneous bursts of laughter from both us, then several more "Oh, nos."

The typographical error was promptly corrected.

CHAPTER 11

ERGONOMICS OF GREETINGS IN THE LAND OF THE RISING SUN

I have always been fascinated by certain Japanese customs—the delicacy, the subtlety, and the finesse in executing the rituals so rich and important in the Japanese culture. None of these traditions and rituals fascinates me more than *o-jigi*, **お辞儀**, the Japanese bowing etiquette most appreciated, yet not well-understood by the outside world.

Several years ago I received my first industrial hygiene assignment in Japan. I was excited but nervous, but it wasn't the language that made me nervous. Our partner company in Japan arranged a fluent bilingual escort to accompany me on my survey. Mr. Ogino, my designated escort, assured me not to worry about communication, transportation, and Japanese customs. I was in good hands. I should confine my worries to my pumps and filters. Thus assured, I concentrated on my field supplies and prepared for the trip.

Despite such assurances, however, I felt terribly unequipped. My worry was about failure of executing that all-important o-jigi, the famous Japanese bow. I had learned that in the Japanese culture, o-jigi is so important that children normally begin learning to bow from a very young age, and companies provide training to their employees in how to execute bows correctly.

The complexity and the importance of this gesture worried me so much that one week before my trip I started reading about o-jigi and began a daily regimen of practice in front of my bathroom mirror. I learned that basic bows are performed with the back in straight posture and the hands at the sides (it can be slightly different for girls and women) and with the eyes down. Multiple parts of the anatomy are involved in this task, but perhaps the middle torso is the most important from an ergonomic standpoint because the waist is where the bow originates. Long and deep bows, reserved generally to convey stronger emotion and respect, can be ergonomically more stressful.

Customizing my greeting

I learned that bows are usually of three types: informal, formal, and very formal. Informal bows are executed at about a 15° angle, or just a tilt over one's head to the front. More formal bows may require a 30° angle. Very formal bows are deeper. Assessing my options, I decided to go for the middle (always a safer course) and focused on mastering the 30° bow.

Duration of a bow turned out to be more complex to tackle. For example, if the other person maintained his or her bow for longer than expected, it was polite to bow again, upon which one may receive another bow in return. This often led to a long exchange of progressively lighter replicates. The angle and the duration were difficult enough. More difficult was customizing the bow according to the social status. Japanese customs dictated that generally an inferior person bowed longer, deeper, and more frequently than the superior. This was a tricky one. Although I was going to meet a fairly high-level executive in this big multinational company, I did not relish the idea of determining the relative status between the two of us. I decided to go democratic and give it equal weighting as far as the duration and repetition of the motions were concerned.

With that decided, the design parameters for my customized bow were set and included:

Angle: *Thirty degree angle from the waist (a semi-formal bow)*
Duration: *Three seconds*
Repetition: *A second bow of shorter duration in quick succession (if the other person was still in the bow posture when I completed my own bow).*
Emergency Response: *A deeper bow (steeper and longer in duration) if an apology was in order and a redeeming bow was needed.*

With the bow parameters selected, all I needed was practice. I quickly realized that practicing in front of the mirror was not enough. For a truly realistic feel, I needed a second person. My always willing partner, my wife, volunteered to be my practice target. That was great, but soon it dawned on me that on the critical day, I would be bowing to a man, so I needed to find a male practice partner.

Normally we don't have another man in our house except when my son David or other manly friends are visiting. No such visitors were anticipated prior to my departure for Tokyo. Out of desperation, I gathered enough courage to ask my neighbor and friend, David Sonne. He was incredulous. He advised my wife to watch out for other signs of odd behavior and seek medical attention if I displayed such crazy behaviors again. David is a medical doctor so he thinks he is qualified to diagnose weird behavior. Nevertheless, being a friend, he obligingly played the role of a surrogate while questioning the sanity of it all throughout the process.

The day of reckoning

Before leaving my hotel room in Tokyo, I reviewed my notes, practiced all the motions in front of the mirror, and, accompanied by Ogino San, traveled to the plant. At the plant, several less critical bows to the junior people (receptionist, plant security, and safety) went flawlessly. Finally, all that hard work seemed to be paying off. I was now brimming with confidence (perhaps overconfidence) to go ahead with the all-important ritual with Mr. Nishikawa, head of the division, who was already waiting in the conference room.

Mr. Nishikawa was a distinguished, wiry, and graceful gentleman with delicate and refined features and a decidedly intellectual look. I walked toward Mr. Nishikawa in an extremely calculated manner, managed an awkward and nervous smile, and executed a well-rehearsed 30° bow of precisely three seconds duration with my eyes facing the floor. When I looked up, I saw that Mr. Nishikawa was still in a bow that could have served as a perfect shot for a poster. Realizing my mistake and remembering that if the other person was still in a bow posture, I must bow again. I managed to quickly execute a second, shorter, but off-balanced, bow. Unfortunately, because of my eagerness to correct my mistake, I overdid it, stumbled forward, and suddenly—WHAM! I head butted Mr. Nishikawa squarely on the forehead with my forehead so hard that there was an audible, somewhat sickening sound.

The damage was done. It was a disaster. The impact left visible bumps on both of our foreheads. Mr. Nishikawa was gracious, despite the injury, in assuring me that everything was fine and that such complicated customs are not really designed for foreigners and that I should not feel bad about it. Well, I knew that my visit, despite my good performance with the technical assignment, was ruined. Ogino San, who, unlike Mr. Nishikawa, had no previous training in disaster recovery, was unable to hide his disappointment. He seemed horrified! He looked at me and actually performed a sympathy bow (a very light bow which I estimated to be well less than 15°).

I wanted to have the opportunity to apologize the Japanese way to Mr. Nishikawa, which is by performing an "apology bow" called *Saikeirei* 最敬礼, literally "very respectful bow." Saikeirei requires a 45° angle and is of long duration. The thought of hari-kari even flashed through my mind.

Anyway, with our disastrous introduction concluded, it was probably for the best that I did not take the opportunity at that time for a second apology bow.

No saikerei 最敬礼.

Bows of apology tend to be deeper and last longer than other types and occur with less frequency. The 45° apology bow is performed with the head lowered, and lasts for at least a count of three, sometimes longer. The frequency, duration, and depth of the bow increases with the sincerity of the apology. In extreme cases, a kneeling bow may be necessary. This bow is sometimes so deep that the forehead touches the floor. I wanted a chance to do saikeirei—although it seemed beyond my physical prowess—to redeem myself but never got that opportunity over the next three days while I was in the plant. Every time Mr. Nishikawa saw me coming from a distance, he gave me a smile but changed course while pretending that that was the direction he intended to go. Of course I knew otherwise.

Multiple root cause analysis

Dan MacLeod, a former colleague and still a friend, is a world-renowned ergonomist and author of several simple-to-understand, hands-on ergonomics books*. He once told me that ergonomics is 90% common sense and only 10% technical. He also told me that many health and safety professionals are qualified to assess ergonomic risks and make ergonomic improvements. All you have to do is to put on your ergonomic glasses. I found this to be valuable advice and since have added an "ergo hat" in my ergo field kit in addition to the "ergo glasses."

Dan's passion for ergonomics is legendary and has earned him the nickname "Billy Graham of Ergonomics" among those close to him. Having once attended a Billy Graham revival a long time ago, I can easily picture Dan at the Los Angeles Coliseum waving at 100,000 followers and declaring: "The Lord will smile at those who wear their ergo glasses, but smite the ones who don't with blindness."

Dan also harps about how ergonomics has as much to do with efficiency and preventing errors as it does with safety. As shown in the following analysis, it was certainly true in my case.

Using Dan's teachings, a simple Multiple Root Cause Analysis (MCRA) and a "fish-bone" diagram, I decided to perform a root cause analysis of the accident. I quickly realized that as in any accident, no matter how simple or straightforward, multiple causes are involved and this one was no different. Because I am a firm believer in technology, I ignored Dan's advice of reliance on common sense alone and went straight to quantitative methods (with Dan's help). I concluded that the following factors were involved:

Posture: While the 30° posture was correct for the initial bow, the angle was too steep for the second attempt. It is hard to judge, but it could have stopped at 45°. Both bows exceeded the guideline of 15° for an optimal second bow. Duration was short, but the consequent error shows the importance of adhering to the guidelines, at least for unskilled people.

Anthropometry: Using Anthropometry tables and knowing that for clearance issues, the statures of tall people are used, I calculated the length of the upper torso for a 95th percentile Indian man at 835 mm and 920 mm for a Japanese man, respectively. Assuming a 45° bow, the standing distance between us should have been 88 cm at an absolute minimum. Let's say 1 meter just to be safe.

Biomechanics: The compression force on the L_5S_1 disc for the two different bows of three seconds can be readily calculated and compared:

Loads	30° bow	45° bow
Peak	1098 Newtons	1803 Newtons
Average	894 Newtons	1425 Newtons
Time-weighted	730 MacLeodians**	1163 MacLeodians**

The strain on the back for the deeper bow is thus 65% higher, a further indication of unacceptable risk for this situation.

Graphed results for the duration of the bows are shown below. Note that the effect of static load on the back for three seconds is included in the analysis, resulting in a slight rise in strain on the back during the bow.

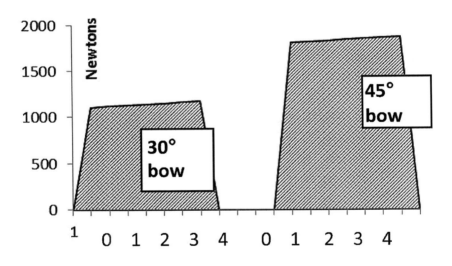

Motion: The second, shorter bow should never have been attempted. In real ergonomics risk assessment terms, two motions executed in quick succession are not likely to result in any cumulative trauma or repetitive stress, but as we have learned from my experience, does have the potential to cause acute injury. In hindsight, I should not have attempted the second bow because I did not have previous training for this maneuver, which required split-second recovery.

Contact Stress: The average head has a mass of approximately 12 kg., but since my head is also filled with a lot of superfluous matter, I upped it to 15 kg. Since the incident occurred on earth, we can estimate acceleration at $\sim 10 m/s^2$, yielding a force of 150 Newtons. Mr. Nishikawa, from my visual evaluation, appeared to have an average-sized head, so the combined force at the point of impact was estimated at 270 Newtons.

Psychosocial Stress: Stress induced by the failure of the first bow resulted in multiple distortions of the redeeming second attempt. The objective measure for stress is concentration of adrenaline in urine; however, this test was not conducted. A WAG estimate for norepinephrine: >80 µg/24hr.

Visibility (inadequate): My eyes were too low (in my clumsy attempt to show humility). This may have obscured my visibility. As we all have learned, potential areas of impact should always be clearly visible. Unfortunately, adding yellow and black safety tape to the foreheads of dignitaries who I planned to meet while in Japan was not considered in my pre-project planning.

Overall, I learned an important lesson in global interaction and basic ergonomics. I have been more successful (maybe fortunate) in my more recent encounters of this kind.

Author's Note: Thanks to my friend Dan Macleod, CPE, for his advice and help in performing a quantitative analysis of the accident.

*The Rules of Work and The Ergonomics Kit for General Industry, both published by Taylor & Francis. Also, a huge online knowledge base of solutions is available at: www.Ergoweb.com.

**Obscure term for the sum of Newton-Seconds, i.e., area under the curve in the graph.

CHAPTER 12

WHERE IS CRISTALINA?

Hazardous chemicals have harsh-sounding and formidable names that are difficult to pronounce and memorize. Such pronunciations are difficult for non-English-speaking people. In my effort to learn basic Portuguese, I discovered that the same chemical names in Portuguese do not sound as intimidating and threatening as they do in English. This is particularly true of asbestos and silica, two of the most prevalent hazardous materials. Every student of industrial hygiene (IH) quickly finds out the importance of these materials: their history, prevalence, and the epidemiology of the various forms of asbestos and of crystalline silica.

Toxic chemicals – Portuguese style

Over the last two decades I have performed scores of IH training programs in front of diverse, multilingual audiences. The hazards of asbestos and crystalline silica form an important part of those discussions.

Students generally take awhile to get used to names like chrysotile, amosite, and crocidolite. The same goes for silica, although saying amorphous silica or crystalline silica may not be perceived as formidable.

While presenting the IH courses in Brazil, it dawned on me that using the Portuguese equivalents can make these harsh terms sound softer and easier to remember. For example, when translated into Portuguese:

Chrysotile: The white asbestos—I call it the blonde one—is *chrisotila* in Portuguese.

Amosite: The brown asbestos—I call it the brown beauty—is *amosita* in Portuguese.

Crocidolite: The blue asbestos—I call it the blue-eyed blonde—is *crocidolita* in Portuguese.

I can't speak for others, but for me these names in Portuguese sound more pleasant and seem easier to pronounce. I have tried this even on English-speaking people. One of the young hygienists from Canada named Kim was so smitten with the name "amosita" that she told me, "Jas, if I have a little girl, I am going to call her 'Amosita'! It is a beautiful name." I agreed with Kim. Amosita is my favorite, too.

"Sounds great, Kim, but make sure to inform the kid at an early stage that amosita is a deadly poison."

"Fine. No daughter of mine will be intimidated by such descriptions," Kim assured me. Knowing Kim, I have no doubts about this.

I am waiting to hear if and when Kim has a baby girl. I will be happy for her. I hope I will receive some credit for her baby's name.

Silica Crystalina

Crystalline silica is perhaps among the most important environmental agents in the history of occupational hygiene. In Portuguese, crystalline silica is known as silica cristalina. When pronounced in Portuguese, it may sound like the name of a beautiful lady, but silica cristalina has caused more occupational illness and misery than perhaps any other chemical agent except asbestos.

I spend much time covering silica and asbestos in my lectures. My presentation in Sao Paulo a couple years ago was no exception. I wanted to educate my audience on the various forms of asbestos and of silica including the toxicology, associated diseases, and the methods to detect and control these hazards that have cut short hundreds of thousands, if not millions, of lives all over the world.

Namorada Gostosa

I was enamored with the Portuguese words for these hazards. For my presentations in Brazil, I memorized whole paragraphs. On one occasion I wanted to introduce to my audience nanoparticles and nanotechnology, which have become hot topics in IH because of the potential health effects of breathing such tiny particles. I decided, this being a brand-new subject, that I should introduce this in Portuguese. I told my audience:

Senhoras e senhores,

Imaginem Materiais:

Menores (smaller)

Mais Fortes (stronger)

Mais Limpos (cleaner)

Mais Inteligentes (smarter)

Senhoras e senhores, isso tem tudo a ver com nanotecnologia.

"This, Ladies and Gentlemen, is what nanotechnology is all about."

The presentation went well. People clapped. Someone in the last row shouted: "Well done." It was Andreia Miguel, the industrial hygienist in the Sao Paulo office.

I am supposed to be Andreia's mentor. It is not clear, however, who is mentoring whom. Andreia is my language teacher, trip advisor, safety guide, etiquette monitor, and my social activities director when I am in Brazil. Ana Paula Medeiros of our office calls Andreia my "Brazilian mom" because Andreia would not hesitate to tell me what to do at any given moment. Strict mom she may be, but she is fairly liberal when it comes to advising me on social interactions.

Andreia does not look like my mom. Truth be told, she does not look like anyone's mom. Well, I better stop here. This story is not about Andreia. I do not want to get into trouble.

Andreia told me that I can be frank and open when talking with Brazilians, including Brazilian women. Especially the Brazilian women! They are independent, free, and opinionated, she told me. They do not get offended as long as you stay two "standard deviations" (SD) below the threshold limit. Also, you could be forgiven and the sky will not fall even if you go over the TLV if you are just an outlier and happen to stumble over the limit unintentionally.

Andreia's coaching emboldened me. I started throwing bolder Portuguese phrases into my presentations, but always stayed two standard deviations below the limit. A more thoughtful friend of mine advised me to stay three SDs below the limit. I could not accept that. Three SDs below the limit would put me in a very conservative category—something I am not.

At this point, I would like to explain that in the engineering profession, the term "standard deviation" is not something kinky. It is an important mathematical (statistical) term related to probability theory. Standard deviation simply means how much variation exists from the average or the expected value. A low standard deviation means that the observations—the data points—are close to the mean or expected value. A high standard deviation indicates that your data points are spread over a large range of values. Needless to say, a large or high standard deviation means out of control or wild behavior that needs to be corrected. Controlling larger than a normal deviation is not easy. I will not attempt to teach anyone IH statistics. I can recommend the real gurus like Dr. John Mulhausen or Dr. Perry Logan to anyone who wants to learn this very important and difficult science.

While teaching the last five-day course in Jundiaí, Brazil, I was having fun sprinkling my presentation with my meager Portuguese vocabulary.

I was talking about the hazards of crystalline silica. I decided to say "silica cristalina" rather than crystalline silica. It was a mistake. The word "cristalina" plunged me into the nostalgic past when I was a young graduate student at the University of Southern California in Los Angeles.

I started thinking about the very first friend I made when I came to America many years ago. Transplanted into the middle of Los Angeles from a small Indian village, I was lost, confused, lonely, and homesick. Then I met Cristalina, a shy young immigrant woman from a country south of the US border. Things were starting to look up. Cristalina and I had a lot in common. We were both new to the area and new to the country. We were trying to learn proper American English. We were conscious of people making fun of our pronunciations although most people understood what we said. Many people thought we were from the same country. They thought we both were Peruvians. Cristalina and I bonded.

Cristalina was everything you could hope for in a girlfriend. She had a quintessential natural radiant look, smoky eyes, and a light brown epidermis. Her bouncy locks made her look flirty. This was too much for a farm boy from Punjab. I set out on an expedited schedule to get to know Cristalina. My research studies could wait. We would talk about our plans after I left Los Angeles. I was thinking of accepting a research fellowship in Ottawa, Canada, that was available to me. Cristalina told me she would follow me to Canada.

I do not know what the top of the world looks like, but I knew then that I was on it.

Where is Cristalina?

One day Cristalina disappeared as suddenly as she had surfaced. I called her room, but no answer. The next day I kept calling, but no one answered. On the third day I showed up at her rundown apartment complex in East Los Angeles and inquired of the old lady in charge of the dilapidated housing complex.

"Are you looking for the young Chicano gal who always wore red dresses? No one knows where she went and she did not even pay the last month's rent. God knows where these people come from and where they go.

Young man, you can find another girlfriend. This place is full of Cristalinas," the old lady assured me.

"I doubt it," I muttered. The lady did not hear me, but it did not matter to her even if she did.

The news was devastating. Where had Cristalina gone? How could she have just disappeared? Had there ever been a Cristalina, or had she been all in my imagination?

Weeks went by without any news. It was as if she had just vanished from this planet. Weeks later I found out that she had been deported to her homeland. They said she was an "undocumented alien," a term immigration officials use to describe foreigners who do not possess valid visa papers or citizenship.

I had heard the term "alien" before. I was called an alien myself until I was granted US citizenship. Still, I had always associated the term with creatures from other planets, not earth. Aliens had weird, non-human features and misshapen or pointy heads. I feel that the term alien at one time may have been derived to distinguish people who were from somewhere else and therefore less desirable. I have no proof of that, but I cannot help thinking that way.

Well, I can assure you that Cristalina was not a conehead or a misshapen creature. She was one of the finest Homo sapiens inhabiting this beautiful planet we call earth.

Back to the industrial hygiene training course in Jundiaí, Brazil—I started talking about the hazards of crystalline silica and could not help hinting to the class that I liked to say silica cristalina over and over again. Andreia finally yelled from the back: "Okay, Jas, just spit it out. Get it over with. Let people know who this Cristalina is so we can move on. Was she your girlfriend? You can say that in Portuguese if you are shy about it."

"Okay, Andreia, Cristalina was my 'namorada, namorada gostosa,' if you will," I replied.

There was a stunned silence in the audience for a few seconds and then laughter. Andreia yelled from the back: "Oh, Jas, you are testing the threshold. Do you realize what you just said? What you said means 'hot girlfriend' in Portuguese. That is why everyone is going crazy."

Andreia immediately called for a break. She was still laughing hysterically.

The exotic Brazilian mom was telling me: "You are supposed to stay within the TLV." I assured Andreia that the term hot girlfriend was not an obscene word. It was well below the two standard deviations from the TLV. My statement was later corroborated by Ana Paula.

"Maybe, Jas, but in Brazil, 'hot girlfriend' could also be interpreted as a 'hired girlfriend,'" Andreia retorted.

Now, every time I do IH training, I intuitively delve into the Cristalina story. When that happens, I develop a lump in my throat. I still wonder . . .

Where is Cristalina?

CHAPTER 13

SIX RULES OF CAMEL SAFETY

People often ask me, "Jas, how did you get started in the safety field?" This is a question I often ask myself as well. I have always been a chemical man. My degrees are in chemistry and more chemistry. I always dreamed of being published in peer-reviewed journals like the *American Journal of Chemistry*—and I was. After finishing graduate school and some post-doctoral chemical research work, my professional colleagues inquired as to what prompted me to go into the safety field. When I say safety, I mean safety in the broader sense where safety, industrial hygiene, and environmental health can be considered one big field. Was there something in my childhood, some spark that pulled me into this field?

The mystery was solved one day when I was telling my kids stories of my childhood. Of all the stories I told my son David and my daughter Monica (neither of whom had yet made it to Dad's birth village) about the palaces, Maharaja's elephants, and tigers, none fascinated them as much as my camel adventures. "How could you enjoy riding such an odd-shaped creature with the scary and unfriendly face?"

Discussion soon turned to safety. "Are you safe when riding such a creature?" "Does it bite?" Not often. "What if you fall, given that you are perched on a hump?" "What about the beast's nasty habits of snorting, sniffling, spitting, and emitting unpleasant hydrocarbons?" I assured my children that despite some of the nasty stories they had heard, camels were not such bad animals and had been unduly maligned. In fact, a camel can be a lovable, amorous creature. At the time I first told these stories, thankfully, my kids were too young to make any sense of this last statement.

Because they anticipated a camel ride in the near future, much of the discussion centered on safety. I assured them that if they followed the six basic rules of camel safety, they had nothing to worry about. I told them to grab their pens or pencils and jot down the rules. You might also find these rules useful as you travel the world and encounter these fascinating creatures as well.

Rule No. 1

Never walk behind a camel

The reason for not walking behind a camel is that the instant the camel senses motion behind it, it wants to kick. A typical camel leg probably weighs around 60 kg. The kick from an adult male camel delivered at a velocity of 100 feet/second will generate a huge force that will result in serious injury and certainly ruin your day.

The second reason not to walk behind the camel is that the air quality in the immediate proximity behind the beast can be (temporarily) highly polluted. I have not confirmed this with GC/MS analysis or with an organic vapor analyzer, but I suspect that it is most likely a cocktail of aliphatic and aromatic hydrocarbons mixed with some aldehydes, esters, and a sprinkling of sulphur and nitrogen compounds, both of the aliphatic and aromatic variety. The very short exposure duration—called "acute exposure" in the health sciences—to such a chemical mixture is not likely to cause "acute injury" or "chronic insult," but may result in a ruined day.

Rule No.2

Always wear fall safety protection

Camels are quite tall, so a camel ride involves spending time in agitated motion at high elevation. You can sustain an injury even if you fall from the animal while it is in a kneeling and relatively stable position. When a camel is standing, the height on top of the saddle, which is perched on top of its hump, will approach the elevation of the second story of a building. For fall protection, a proper saddle (often ignored) is a must, and the saddle must be properly positioned around the hump. Note that I have experience only with the one-hump animals. I am sure the beasts with two humps have different requirements and safe operating procedures.

Standard safety gear, including a belt, harness, and helmet, are always recommended.

Rule No. 3

Wear eye and foot protection

Eye protection with UV absorption lenses treated for glare are recommended, especially in arid or tropical climates where this transportation mode is most often used.

Safety shoes with cushioned heels (to soften the impact upon landing from a fall) are preferred. This requirement can be waived if you are in a sandy area where there is no hard surface that you could land on.

I also recommend cushioned stirrups to avoid unintended bumping of the beast's belly with your shoes, triggering uncontrolled flight by the animal. All horse riders know well that hitting the animal's belly with the heel of your boot is the signal for the beast to speed up. Camels recognize this international signal for "accelerate."

Rule No. 4

Be nice to your camel

A camel is generally obedient but unpredictable. Remember, the animal has an amazing memory; it never forgets. This means that if you are nice to it, it will be nice to you and will reward you with dependability and a level of loyalty rarely found in humans. If, however, you mistreat a camel, you'd better watch out! It will not forget and will never—I mean never—miss an opportunity to get even. Payback usually comes when you least expect it. Stories abound in my village about a farmer who roughed up his camel one winter (camels sometimes get unruly and uncooperative during colder months— see rule no. 6) in a clumsy effort to subdue the beast.

After roughing up the beast, the farmer barely escaped with his life on several occasions. The animal never forgot and never forgave the man for such brutal and abusive behavior. Rumor has it that over a four-year period the camel tried to snuff him out in a variety of ways including kicking him and once sneaking up behind the unsuspecting farmer and biting the back of his neck.

One hot summer afternoon the farmer decided to rest under the shade of a big pipal tree. Pipal trees are revered in India for their thick shade during the hot summer months. When the man was half sleep, his camel, bent on vengeance, quietly sneaked into the shade beneath the same pipal tree and sat squarely on the farmer's chest. Well, you can imagine a 1500-pound beast sitting on you. The barrel-chested farmer, with the agility of a gazelle and the strength of an Olympic wrestler, successfully dislodged the beast from his torso. Soon afterward, the farmer sold his prized possession at the annual livestock auction in the nearby village. If only he had applied rule no.4 he would not have lost his valuable asset and not almost lost his life.

Rule No. 5

Always keep a minimum two-camel distance between animals

A camel is unpredictable and sometimes outright nasty. Even in the same family, camels do not get along very well. They are always itching to fight, given the chance. The saving grace is that serious injuries from such camel-versus-camel fights are rare because the animals do not have sharp claws, horns, or fangs that could result in grievous injury or death.

Nonetheless, failure to maintain the minimum two-camel distance may result in the two beasts contacting each other and possibly hurting each other.

It seems to me that the two-camel separation rule may have been developed by farmers in Asia (or perhaps in Africa, but I can only speak for the Asian camel since I have little experience with their African kin) and was later adopted by the so-called transportation safety experts in western countries.

Rule No. 6

Never tether a male camel within eyesight of a female camel during the winter

Male camels can sometimes be very active and excitable in winter. Consider reviewing rule no. 4 above. The fact is they can get very amorous. This will no doubt sound odd to people not well versed in camel behavior psychology. Those who know the creature know that it is a very romantic beast (unduly maligned and ridiculed, as I said before).

When a male camel sees or smells an attractive female camel, it wants to get to know her but unfortunately is usually prevented from physical contact by its short leash. The consequence of this is predictable; the animal goes into a deep depression, whines all the time (camel whining can be very annoying), and just wastes away.

Rule No. 6a

Always keep camel condoms around during the winter. According to rule no. 6, male camels can get very amorous during their winter mating season. Condoms are used to help prevent unwanted pregnancies and prevent camel venereal diseases. In all seriousness, camel condoms were discovered centuries ago by Saharan traders.

Author's Note: Like everybody else, I have envisioned my retirement and what I will do once I am off the corporate payroll. My latest thought is to conduct tours of India for Western tourists in cooperation with my daughter Monica Haley, who has become an expert on India. The main feature of our tours could be camel safaris in the deserts of Punjab and Rajasthan. I figure, all other things being equal, no other tour operator could match the qualifications and benefits offered by the Singh Travel Team. I think of my qualities— a professional safety expert, a board-certified industrial hygienist, and a PhD tour operator. I don't want to boast, but what other tour operator has similar safety credentials?

CHAPTER 14

ETHNIC DIVERSITY

America is a melting pot, an ethnic mosaic of nationalities, cultures, customs, and languages. This melting pot has produced a unique global community and spawned wonderful new traditions including music, fashion, cuisine, and the arts. Sometimes the warmth from the melting pot causes heat stress to some individuals, leading to societal rifts. The good news is that professional industrial hygienists are around to help alleviate this heat stress. Ethnic diversity is what has made America great and has contributed to comedy, humor, and belly laughs. Enjoyment of all of these attributes is recommended for a long and happy life.

The Puerto Rican next door

Early in my career, I was in Houston, Texas, doing my usual job of chasing battery-operated air-sampling pumps at a chemical plant that made resin composites, when an urgent call came in from a major insurance brokerage house, interrupting my work. Someone had an emergency.

The caller was Josh Atkins, an account executive with a major insurance broker in New York. We both worked for the same company but in different divisions. I had assisted Josh on many safety and industrial hygiene projects for his favorite accounts. During those occasions Josh and I became friends. He was one of the best 'BS-ers' in the business! So good that a sign inside his office read, "I am a good bull 'sh!tter' myself, but occasionally I like listening to an expert, so please carry on."

The first time I went to see him I wanted to tell him about my own exploits, but the sign on his desk disarmed me, rendering me ineffective. I learned a lot from Josh. He taught me colloquialisms (American slang), enhancing my American vocabulary mostly with words that I cannot use in polite company.

Josh's client, the insured, had an emergency at their plant in Puerto Rico. The plant made costume jewelry and had suffered a toxic release leading to the evacuation of all the employees. The toxic fumes sent nearly two hundred employees to seek emergency treatment at local clinics and hospitals. At least thirty workers had to be flown to the large hospital in San Juan. Two of the employees were in serious condition, suffering from respiratory distress. Apparently the fumes were from chlorine gas.

"Now you know as much as I do," Josh told me. "Jas, please wrap up what you are doing. We can staff your Houston project with someone junior. They are so panicked over there in Mayaguez that they need someone senior, preferably someone they can call doctor," Josh emphasized.

"They can call me a pimp as far as I am concerned, but please tell them that I am not a medical doctor. I am a chemical doctor."

"Excellent," Josh responded. "A chemical doctor is what they need at this stage. They have medical doctors coming out of their . . ." He hesitated and said, "Ears."

"Now listen, Jas, if you catch the next flight to San Juan—three hours from now—you will reach San Juan by midnight. Someone will book a room for you at the airport hotel where you can grab a couple of hours of sleep. In the morning you can catch the first puddle jumper to Mayaguez, the airport on the western coast of Puerto Rico. The airport is only thirteen miles from the Mirabella plant." Josh summarized all this in one breath; later he told me he could hold his breath for incredibly long periods. Josh explained much more about Mirabella, his prime insurance account. He wanted to dazzle his client with our speedy response, which required my extraordinary effort, of course.

"Understood, but what will I do when I reach Mayaguez after travelling on that deathtrap of a flying machine with two little flyswatters disguised as propellers?" I was referring to the twin-engine Cessna plane I was to fly on.

"How will I get from the airport to the plant thirteen miles away? I don't speak the language, and I have never been to Puerto Rico before," I questioned.

"Jazz man, you will have no trouble. You are a well-traveled man. You look like a Puerto Rican, and I know you can bullsh!t your way using the three Spanish words in your vocabulary even though two of those three words cannot be said in polite company."

I thanked Josh for his overwhelming confidence in me but asked for the phone number of the Mirabella plant manager. This manager had just come to the plant from Chicago and was new to the assignment.

The next morning I boarded the 6:30 a.m. flight from San Juan to Mayaguez. As I had feared, it was a small twin-engine Cessna, the same kind of airplane that I had flown with my fellow industrial hygienist, Charles Barkley, on dozens of industrial hygiene surveys in small towns all across the USA.

The twin-engine Cessna had only eight passengers—two ladies with three toddlers, and an old man who constantly coughed; I feared for his health and longevity. Then there was a teenage boy and me. I was wearing business clothes. Later on I thought I might as well have hung a placard on me saying, "Excuse me. I am weird. I don't belong here." A two-piece suit was a bad choice for my travel attire.

Before leaving San Juan, I had called the Mirabella plant manager at his home to advise him of my arrival at the Mayaguez airport and asked him to send someone to pick me up. He assured me that his safety manager, Louis Valez, would be there to meet me and take me to the plant. Moreover, Louis would explain the details of the incident and the reason I was coming to Mayaguez. He added, "Louis speaks good English."

When we landed, as the eight of us emerged one-by-one from the Cessna, I was expecting a loud greeting from a Spanish voice saying, "Hey, doc, welcome to Puerto Rico. Is that all your luggage? Let us get the show on the road. The big boss is waiting."

I heard no such greeting. Fifteen or more minutes passed and no sign of Louis. I began to get nervous. Then I heard the intercom. "Señor Singh, please report to the office. Someone is waiting for you."

That someone was Louis. Louis gave me a 30 PSI (pounds per square inch—a measure of force) handshake and said, "Doc, welcome to Puerto Rico. Man, I am glad you are here. I was getting worried. I was here fifteen minutes before the flight. I saw the few passengers come out. There were a couple of ladies, a few kiddies, and an old dude who I thought would croak before he hit the tarmac. Then there was another guy dressed up like a pimp (me), but I did not pay any attention to him."

"Louis, I am surprised. I mean, how could you miss me coming out of that little airplane? There were two ladies, three bambinos, a teenage kid, a sick old man, and me, the pimp on the flight. I should have stood out like a sore thumb among that motley bunch!" I challenged him.

"You did stand out, man. I apologize for calling you a pimp. I feel stupid. To tell you the truth, I saw you. The boss had told me that I was supposed to pick up a Chinese dude named 'Sing.' No one else had any more information about you—not your first name, your photo, your height, your skin color, nothing. So I was searching for a Chinese face. I saw you with my own eyes. You looked just like another crazy Puerto Rican, only dressed up like a pimp."

"Thanks, Louis." I said. "That is what I call getting rid of all formalities and getting down to business. I like that. I am not into formalities either. Let me tell you, I am flattered to know that I look like the typical crazy Puerto Rican dude, even when I am dressed up as a pimp. This means I am accepted, I am in!" Louis and I shook hands again, and we were quickly on our way to the Mirabella plant.

We bonded instantly. Louis gave me the summary of events during the thirty-minute ride and also told me: "It is good that you are here. We had a chlorine leak from a big cylinder tied to a little pole right below the air-conditioning intake. We use the chlorine to disinfect drinking water in a hurry if the drinking water reservoir gets contaminated during a hurricane. Over the years the chlorine corroded the metal and broke through with a big hiss. The cylinder, being under pressure, released the gas in a big cloud that got sucked into the air-conditioning system and gagged the more than two hundred employees. Thank the Lord everyone is safe, but can you imagine the stupid genius who installed the chlorine tank in a place where it would choke the maximum number of people?"

"You are correct, Louis. I don't want to say the person who did this was stupid, but on the other hand, he was no valedictorian either."

Louis laughed and said, "I think we will get along okay." He then gave me a tour of the whole building. No people were inside. The workers had refused to enter until an expert proved to them that the toxic gas was gone. I was sure no toxic gas remained. The gas could not hide inside the building. The air-conditioning would have flushed it out of the building a long time ago.

Louis and I showed the numbers to the plant manager. He immediately called a meeting with the rank and file workers who were hanging around outside the building. The employees were anxious to get back to work, as was the management. The plant manager instructed Louis to tell the people that the company had hired the best toxic gas expert in America—an expert who had many years of experience working in this field. The expert would show them the results of the testing.

Brimming with confidence, Louis introduced me to the crowd. He stated: (later translated into English): "Senoras e senores, this is Dr. Jas, one of the top experts in the United States. Jas has brought from America the best gas detection equipment that money can buy." (This statement was not entirely true. All I could muster in the short time was an old gas detection meter and some rudimentary grab sampling colorimeter detector tubes.)

"Jas will show you the results, and I will translate for you what the doc is saying,"
Louis could not resist saying. "Doc knows some Spanish too, but not the kind you can use in front of the ladies." The crowd laughed. He then thanked everyone and urged them to listen.

He then added, "Please make sure you meet the doctor and shake hands with him. You will like him. He is very funny." I thought that was unnecessary, but it seemed to release the tension.

Soon we achieved a successful conclusion. I had one extra day before I had to go back to the USA. On that day, Louis informed me that we would go out on the town to celebrate. He outlined the whole evening for me this way: "We will leave the plant at five p.m., drive by my house so you can drop your bag there, and tell my old lady that I will be a little late." (That meant two a.m.). "Then we will stop at the Mayaguez Hilton for a drink. The Hilton is the best place around if you want to have fun. The piña coladas are the best in the world, and the local chicks are always hanging around the bar. Puerto Rican chicks are the best in the world, but I will let you tell me what you think at dinner. We will have dinner at my favorite place. After that we will go to the shore to watch the water 'glow' where the little bugs in the water light up the water surface like millions of little Christmas lights or like stars in the sky. After that, maybe we can have a couple more drinks back at the Hilton bar where you are staying. Piña coladas will still be there and if this is your lucky day, so will the chicks."

"It is already my lucky day even if the chicks are not there," I responded. "You are a great host and I am in your hands. Just make sure I am in safe hands," I told him.

"They don't call me the safety manager for nothing." He quickly dismissed my safety concerns, if I had any.

The highlight of my trip to Puerto Rico was yet to come. We were close to one of the most spectacular "bioluminescent bays" in the world. The unique bay in the area was known to contain close to one million single-celled bioluminescent dinoflagellates per gallon of water. These half-plant, half-animal organisms emit a flash of bluish light when agitated at night. The high concentration of these creatures can create enough light by which to read a book.

The mysterious blue-green light is created by microorganisms that thrive in an environment uniquely suited to their needs. Observing the phenomenon from kayaks on a balmy night has been described as a magical experience. The night was beautiful and the luminescent phenomenon was awesome—a spectacular display of nature! I had heard about such, but I had never witnessed anything like this.

When we came ashore after our kayak trip, Louis started talking to the boatmen. He said in Spanish, "This is my friend, Doc Jas. Can you guess what country he is from?" Half a dozen boatmen started staring intently at me and finally our boatman said confidently, "He is Peruvaño." Others nodded in agreement and were excited in the expectation that as I was a Peruvaño, I could converse with them in Spanish.

I shook my head to indicate no. Louis came to my rescue and explained to the boatmen that I was not a Peruvaño, but an Indiano Americano. Everyone smiled and endorsed my ethnic description. It sounded good to them.

The next day Louis took me to the airport. The little Cessna was ready to go. Louis gave me a warm Latin hug, and as I was walking onto the tarmac, he slipped a bottle of premium Puerto Rican Caliche Rum into my carry-on bag (the Cessna had no restrictions on carrying alcohol).

I was sad to leave a friend I had known only for three days. I never saw Louis again despite many promises to come back and spend much more time.

Funny Peruvian

The American Industrial Hygiene Association's annual conference in 2008 was held in Minneapolis, the home city of Greg Beckstrom, my friend and my "chief editor."

My colleagues always wanted me to show up at this event; many people we knew in the industry sought me out at our exhibition booth. They also threatened to create a life-sized cutout photograph of me wearing a nice suit, a big smile, and a pair of dark sunglasses. I refused this generous offer because every time I thought about it, I was reminded of Louis Valez from Puerto Rico once mistaking me for a pimp.

Tuesday was a busy day at our display booth. On that day Brian and Shamini were tending the booth along with me. Lunchtime arrived on what was an exceptionally warm and sunny day in the Twin Cities. Many outdoor eateries were doing a brisk business. As luck would have it, a well-dressed young man walked up to Brian, gave him the 30-PSI handshake, and said, "Do you have any plans for lunch? Lots of outdoor places are open on a rare day like this in Minneapolis."

"No, I do not have any plans. I was just discussing lunch with my two colleagues here." Brian introduced Shamini and me to the man, who appeared to be in his thirties, and wore a dark suit and a suitably starched white shirt. He was very white.

"Then let's go. Your colleagues are invited. I know the town. Let us go to my favorite place not too far from here."

Shamini and I looked at each other and winked. *There is a God after all,* we thought (not that we doubted it). I thought: *A glass of chardonnay is in my future.* I assumed that the man had a generous travel allowance. Our benefactor was an account executive for the company that insures our firm. We were his clients. This relationship was good enough for at least one glass of chardonnay.

The restaurant was nice. We settled down in comfortable chairs in a sunny spot and introduced ourselves. After initial niceties, the insurance man wanted to know more about us. He had noticed a striking ethnic phenomenon around the restaurant. He was especially curious about the ethnic mix represented at our small table. He knew his own ethnicity (one would hope), but he never told us. First I thought he was from Scandinavia, maybe Iceland, but later I found out he was of Russian descent. I do not remember his name, but let us call him 'Boris.'

The four of us around the table comprised the wonderful ethnic mix so well noted in the United States of America. Looking around the table, I figured we had a German American, an Indian American (not an American Indian), a Malaysian Canadian, and a Caucasian Russian American.

Boris could not help saying: "America isn't called the melting pot for nothing. I am curious to know where everyone is from." He first focused on Shamini (not a big surprise). I suspected he was intrigued about Shamini for more than her ethnicity.

"So Shamini (which he pronounced 'Shāme ā nē'), what is your origin? You have the most interesting profile I have ever come across" (flattery does work). Shamini felt compelled to correct his pronunciation of her name, which is "Shăh mă nē." Then she explained how she was born in Malaysia of Indian parents and now practiced industrial hygiene in Calgary, Alberta, Canada. Boris clapped his hands and said: "Awesome!" He never took his eyes from Shamini's face.

Boris then turned to Brian, smiled, and said, "Well, Brian, I don't have to guess about you from your last name (which is "Senefelder"). What part of Germany is your family from? I believe my parents originally came from Germany, too."

Brian produced his characteristically patient smile and said, "Yes, that is one advantage of having a last name like mine. People don't have to strain to guess my heritage."

Boris was saving the best for last. "Jas, I have been wracking my brain to determine your heritage. You could be Mexican or Guatemalan, a Sicilian, or maybe even Egyptian. I have seen several movies starring Omar Sharif. If you sported a decent mustache, you could look like Omar Sharif." He never mentioned Indian or Pakistani as a possibility for my heritage.

"Why don't you take a guess at my nationality? See how close you can come to the right answer," I responded. I thought this would be fun.

"Well, I am sure you are from Latin America. You could be from Mexico or one of the Central American countries or even farther south. You could be a Bolivian or a Peruvian or even a Brazilian Mulatto." He hesitated a bit at that one.

"You are hovering close to the target, Boris. I am Peruvian."

"I knew it!" In his excitement he almost jumped out of his chair.

Boris congratulated himself for the bull's-eye hit. He could not hide his glee.

He continued, "So tell us a bit about your childhood in the old country. What brought you to good ol' America? Your parents must be proud. I hear things are not that great in the old country. You must thank your stars every day!" Boris exclaimed.

"Yes, I do, but many days it is cloudy and I cannot find any stars to thank," I told him.

Boris let out a hearty laugh. He looked at Brian, then pointed at me and said to Brian, "This guy is funny. He has adapted well to his new country." Brian shook his head in agreement.

Before I told Boris about my childhood in the old country, I noticed some movement under the table. Brian and Shamini were kicking each other while trying (unsuccessfully) to conceal their smirks. They were going wild. If I had not known the two so well, I would have thought that some hanky panky was going on between the two. I knew better.

"Boris, I don't really know how things are back in Peru. I left when I was a young boy. I have not been back for many years but I must go back some-day." Then, to fend off one big embarrassing situation, I said: "You know, I can't even speak my mother tongue anymore. I am not proud of it and actually I am embarrassed. This is one reason that has kept me from visiting my homeland."

I wanted to preempt Boris on the language issue. Many Americans, especially those living in the Southwestern US, can surprise you sometimes with their working knowledge of the Spanish language. I have been caught in this trap before where a simple, innocent lie engulfed me in a spider's web.

I was aware of another small tremor under the table. Shamini and Brian had another footsie exercise. They were at the point of breaking with all the Peruvian expat details. Finally Brian could not take it anymore and excused himself to go to the washroom. Left alone to deal with the mounting BS, poor Shamini had no one to kick under the table. You cannot play footsie with your own two feet.

I wanted this discussion to end, so I summed it up for Boris. "You know, Boris, I need to catch up on my Spanish and then visit the place of my birth as soon as possible, reclaim my heritage, and also try to find my other countrymen. I know there are a lot of Peruvians in the US. If we meet next year at the AIHA conference, let us have lunch again. You will be impressed with my progress I will make on my Spanish in such a short time."

"I don't doubt it." Boris paid the bill. We shook hands and Brian, Shamini, and I headed back to our booth. Before Boris parted, he pulled Brian aside and I heard him say, "The Peruvian fellow is real funny. Make sure to bring him again next year and let us do lunch again." Brian assured him that the Peruvian fellow never missed this conference or a free lunch.

As soon as we moved a safe distance from Boris, we burst out in laughter. We mused: "How could a college graduate with years of experience dealing with business people believe this? How many Latinos could he have come across named, 'Singh'?" This was unbelievable. I thought, *Perhaps US schools should have a compulsory course in ethnic or cultural diversity.* This understanding is important no matter if you are a geology student or a philosophy major. You need to know about other lands and other cultures.

We entered the exhibit hall and resumed our booth duty, still laughing. Our other colleagues at the booth looked at us in a kind of weird manner. Finally, Charlie said: "What were you guys smoking during lunch? You know it does not set a good example to the younger people." I agreed it did not. I had every intention of telling Boris the truth, but I never found the opportunity. The BS had spread too far and too fast!

Snobbish Latino

Years ago I was walking down Vermont Boulevard near the University of Southern California campus where I was a postdoctoral fellow in the chemistry department when a man caught up to me. He was eager to strike up a conversion.

"Hola, señor. Qué tal?"

I understood the hola part but was not sure of the rest.

"Pardon me?" I said.

He repeated; "Hola, señor. Qué tal?" This time he raised the decibel level. Maybe he thought I was hearing impaired.

Then it occurred to me that he was probably saying "How are you," but I was not sure. Better to be safe than sorry.

"I am sorry, sir, but I do not speak Spanish," I tried to apologize.

"You do not speak Spanish? Your own mother tongue? This is a pity."
He was irritated and switched to English, which he spoke very well.

He then said, "I know you are a Latino. It is such a shame. People come to America and they become snobs. They deny their own mother tongue. They deny their own heritage . . ." Although he was still going on, I felt compelled to interrupt him.

"But sir, I am not Latino. I am from India. I have never been to South America. Someday I would love to learn Spanish. It is such a beautiful language. I love Latin music. I even had a Latino girl friend at one time. Her name was Cristalina. Do you know her?"

The man looked at me. His face deflated. He was not a rude person. He started apologizing profusely and said: "Pardon me, señor. I am so sorry. I was rude. I was sure you were Mexican because you look just like one of us."

I told him there was no need to apologize and that his saying that I was just like one of them was very friendly and gracious. We shook hands and parted.

Later I kept thinking about the sentence, "You are just like one of us." I thought more people in the world should learn this one English sentence. I had read somewhere that the most-recognized English word in the world is "okay." Even the most remote inhabitants on this planet understand this word.

I said to myself, 'Wouldn't it be wonderful if we taught the world one English sentence to make people feel welcome in a foreign land. Just one short English phrase:

You are just like one of us.

CHAPTER 15

GREAT AMERICAN
SALVAGE OPERATION

On January 17, 1994, the San Fernando Valley of California was shaken by a 6.7 magnitude earthquake. The damage was extensive—far greater than what would have been expected from a 6.7 magnitude trembler. Apart from the tragic loss of life and extensive property damage, merchandise worth millions of dollars was coated with plaster, insulation, paint chips, dirt, and dust particles buried under the collapsed roofs of specialty stores, big-box stores, shopping malls, and fashion houses in the Santa Monica area.

Toxic fallout

As soon as the immediate health concerns subsided, attention shifted to property damage. Hundreds of buildings lay in ruins or were severely damaged, bearing yellow or red tags depending on the extent of damage and probability of pending collapse. One familiar store had tons of expensive merchandise buried under a collapsed roof. The merchandise had a uniform brownish white layer of a very fine powder that appeared to be a combination of gypsum, cement, and white paint. Someone suggested it could be asbestos since the building was known to contain asbestos fireproofing and acoustical insulation. Attention immediately centered on what to do with the millions of dollars of merchandise possibly contaminated with the deadly fiber. Stories of asbestos liability lawsuits abounded. Potential liabilities from the use of contaminated merchandise could be high even in the absence of proof of health damage. An extra cautious consultant in town recommended to the company that the stock damaged by the cancer-causing white powder be treated as hazardous waste, double bagged in plastic, and buried in a landfill away from the prying eyes of the public. The value of the damaged stock from this store alone was estimated at $8 million dollars.

Indecent proposal

My friend and colleague Charles Barkley, a recognized expert on asbestos exposure and liability, was outraged. When he heard the news about the proposed disposal of this pricy merchandise in a landfill, he was incredulous. He immediately called me and criticized this goofy decision in his own colorful narrative, which I am unable to recall and even if I could, I pledge not to repeat here. He described the decision to bury millions of dollars of perfectly good merchandise as "pathetic."

Charles was a born entrepreneur. He was also a recognized expert in asbestos management, including risk assessment and reconstruction of past worker exposure to asbestos. Charlie was particularly adept at assessing damage and resulting potential liabilities from the use of asbestos-containing materials. Part of what he told me was that it was moronic to throw the perfectly good merchandise into an engineered pit—merchandise that could be decontaminated, tested, proven free of contamination, and sold on the secondary market, yielding perhaps up to 40 cents on the dollar for the owners or their insurance company. It sounded good to me. I am a believer in technology and revenue. Merchandise could be cleaned free of fiber contamination. This was not a material that would soil or seep into things. Fibers would settle on the surface. Much of the merchandise was in plastic covers or packed in cardboard boxes that were still intact. Asbestos could not permeate by any known pathway. We had the technology to clean the merchandise to a satisfactory level and prove it with statistically valid electron microscopy techniques.

"Let us propose to the owners that we will decontaminate the damaged goods under environmentally safe conditions, test the cleaned-up merchandise, and provide statistical proof of purity by a certified industrial hygienist," Charlie suggested.

"We will check with our lawyers, explain what we intend to do, and if approved, let us go for it. Our fee will be fantastic because we will propose to be paid a certain percentage of the resold goods," Charlie assured me. Once again, it sounded good to me.

Our proposal was accepted in a flash. Papers were signed. Charlie convinced the owners that they would recover between $3 and $4 million dollars' worth of merchandise, and our fee would only be 15%.

Planning the heist

Charlie rented an abandoned warehouse in Anaheim and had the damaged goods trucked to the building. The warehouse had been out of use for several years and had to be thoroughly cleaned, vacuumed, and aired out before bringing in the merchandise and the people. Lighting and ventilation were revamped prior to the decontamination start-up.

Charlie went on assembling the team. The cleanup crew consisted of four employees from the nearby Orange County office. One of them would act as the crew supervisor and be responsible for implementing the health and safety procedures that included the standard asbestos containment procedures, personal protection equipment, personal hygiene, and waste management, among others. This motley crew was supplemented by importing two more technicians from our East Coast office in New Jersey. Two of the four Orange County employees were recent immigrants from the Philippines. Both had some experience in the cleanup of hazardous wastes. One of them was Rico Hernandez, who became a good friend of mine because of his diligent work ethic and his humorous nature. The rest of them were health and safety technicians who had never performed (neither were they expected to) actual manual labor. All of them adapted to their new profession quickly. Financial incentives were a big factor.

Solid proof of decontamination was the most critical element in this novel experiment if the salvaged merchandise was to be sold on the secondary market. This task was delegated to me. Charlie, the supreme commander, appointed me as his "quality assurance director" without even asking me first. He assumed it would be okay because he believed (mistakenly) that I was a publicity monger always looking for more titles, even if frivolous ones. He did not elaborate why he appointed me as the QA director except that he told me that the letters (PhD, CIH) after my name would impress the jurors if it ever came to a trial. I was not sure if this was a compliment.

Charlie was the man. He was the contract manager, as well as the supreme commander. He had all the requisites for this unusual assignment. He was a Vietnam veteran, an accomplished pilot, a race car driver, a great storyteller, and a taskmaster. I had flown with him on many occasions all over the United States, working on previous assignments. There were many exciting (if you could call those exciting) moments flying with Charlie, including an emergency landing in Baton Rouge in the midst of a thunder storm.

Charlie was an Easterner who lived in the South for a major part of his life, and he had thoroughly assimilated the Southern styles. One of the highlights of my frequent trips to his home city had always been dinner with Charlie and his girlfriend. These evenings usually ended very late.

The lean mean cleaning machine

After much preparation, decontamination started in earnest. Soon the old warehouse in Anaheim was filled with expensive designer garments, shoes, handbags, and more. Expensive bridal gowns wore tags that read: David Tutera, Olga Cassini, Marliana Liana, Demetrio, and Simon Caravalli. Having been born in a dusty little village in Punjab, India, those labels did not mean much to me, but when I mentioned them to the ladies back in my Orange County office, I heard more than one gasp. Some of the gowns were worth thousands of dollars.

Items that caught my eye were the men's designer colognes. Armani, Versace, Lacosta, Calvin Klein, and Paco Rabanne (my favorite) labels in shapely bottles lay undamaged in their designer packages. I was convinced that all the items could be decontaminated. Our testing protocol required that a randomly selected piece of decontaminated merchandise be tested by electron microscopy for any evidence of asbestos particles, and if no trace of the toxic fiber was found, the lot was stamped "approved" after assuring statistical proof.

Our motley but determined crew was ready. Charlie gathered the gang and gave them his usual pep talk. He told us to think of the uniqueness of this project and the big financial rewards we stood to reap. Moreover, it could be some kind of a breakthrough in salvaging precious commodities and preserving resources. He made it sound like this was a patriotic act we needed to take seriously. The patriotic part particularly impressed the two recent immigrants among us.

With the silent cheer of "Please leave us alone," we went to work. We had only three-and-a-half weeks to finish it all. Delays would incur severe penalty.

The 'one molecule' curse

Charlie and I agreed that we needed ironclad quality assurance protocols. Before we would tag cleaned-up merchandise as "approved," we wanted to make sure that our tests were authentic, state-of-the-art, and carried proof of no detectable asbestos structures with statistical reliability.

Both Charlie and I were cognizant of the liability issues attached to this endeavor, especially in the asbestos gold rush environment where many charlatans were looking at making a quick buck, claiming health damage from exposure to asbestos whether their work history supported any evidence of exposure or not and whether there was any evidence of health impact. We conjured up scenarios where it might be possible that someday someone wearing our salvaged garment would sneeze several times in a row and scream: "Oh my God, this is the result of wearing that bargain designer dress I bought at the resale market."

We wanted to be prepared. Our testing protocol involved surface sampling of thoroughly HEPA (high efficiency particulate arrestor)-vacuumed clothes, and if the decontaminated specimen passed our stringent test, only then would it be added to the approved pile and a copy of the test certificate attached. This thorough process resulted in setting aside a number of expensive items that looked perfect but were slightly damaged by an adhesive stain from the sticky tape used to "lift" dust and fibers from the garment. Our agreement with the sponsor was that we could keep the damaged samples for our personal use or give them to the poor (who included our own technicians, the cleanup supervisors, and the QA director.)

We were confident that our methods would help our sponsor ward off any phony claims (yes, with our thoroughness, a claim had to be phony). We were equipped to go to battle with those pseudo scientists, the so called "one molecule" theorists who maintained that we must get rid of scourges like asbestos to a one-molecule level because even a single carcinogenic molecule could trigger cellular damage. Maybe, but in practical terms both Charlie and I believed that this was silly; we still do. I could delve further into this, but it was not funny.

Most eligible bachelor in Manila

Rico Hernandez was a recent immigrant to California from the Philippines. He had joined our company as an asbestos technician to accompany more experienced hygienists on field surveys. Rico was a good-looking man, and he had already assimilated the modern California punk style of the day. Walking on a street you could not distinguish him from the spoiled Orange County rich kids. The reality, however, was very different. Like many newly arrived immigrants to the land of nuts and flakes (California), his financial situation was not to be envied. He was sharing a grungy little apartment in Anaheim with one of his classmates from the old country. His English was awkward and halting. Rico made up for these limitations, however, with his good nature, broad smile, and an easygoing friendly style. I liked Rico and we soon became buddies. Charlie liked him, too. Having spent most of his life in the South, Charlie had developed a keen eye for spotting less demanding, loyal, and obedient helpers. Rico was his man.

The salvage project was a godsend for Rico. With sixteen-hour shifts, the second half of which was at a 50% premium, Rico could pay off some of his debts and make his long-planned trip back to Manila to visit family and friends. Moreover, he could take home some very expensive designer merchandise to give as gifts—merchandise that had been cleaned but was not approved for resale due to minor flaws or damage from our testing. He was very excited.

I told Rico that once he returned to Manila, he would be the most desired bachelor in his neighborhood. He could show a prospective bride the $4,000 Olga Cassini bridal gown he had salvaged with his own hands and as a bonus throw in a box of coveted designer perfume. He had a choice of Christian Dior, Elizabeth Arden, or Ok One by Calvin Klein. I could even give him the bottle of Hugo Red that I had randomly selected for testing; this was permitted under our agreement.

I pulled Rico aside and said, "Look, man, what woman will not fall for you once you flash that Olga Cassini gown while gently rubbing a dab of the Hugo Red on the back of her hand just like they do at the Macy's at the big Orange County Mall?" I continued, "Your whole place will smell like a perfumery on Paris's Rue du Faubourg Saint-Honoré or Milan's Via Montenapoleone!"

Rico was quick witted despite his faulty English. He interrupted me. "Yes, Jas, and my place will smell like a French whorehouse; maybe my whole neighborhood will smell like that." Then he quickly added, "Mind you, I have never been to a French whorehouse or any other one. I could never afford it. I just read about it."

"Well, my friend, neither have I, even though I can afford it."

Hard-earned success

The project ended on schedule as promised. Cleaned-up merchandise was neatly stacked inside the rented warehouse under marked categories: bridal gowns, women's clothing, men's wear, children's clothing, designer perfumes, men's colognes, designer bags, purses, etc. We were assisted in this inventory by one of the store employees. Yellow tags of "approved for sale" stickers hung all over One of the stacks was labeled "Damaged Samples, Not for Sale." Our motley crew, however, was free to "adopt" these if they so desired. These were perfectly good materials. We all had our own shopping bags ready to haul away the loot.

Celebration

The time had come to celebrate and say goodbye to colleagues from New Jersey who had pitched in on this patriotic effort, three-and-a-half weeks in a row away from their families and friends.

Charlie organized a grand feast in his characteristic style. Dinner was held at the Orange Hill Top restaurant that featured cascading waterfalls and Koi ponds, as well as an English-style lobby with a fireplace. On the day of our celebration, the restaurant provided a rare smog-free panoramic view of the valley below. We wore our best clothes, and everyone sported a HEPA-filtered salvaged souvenir of one kind or another. I wore a designer Aloha shirt from Tommy Bahama® that I had personally tested and certified. I had successfully removed the brownish stain left by the sticky sampling tape from the test sample. My designer Aloha shirt did not look defective in any sense of the word.

Charlie thanked everyone. To the thunderous applause of the clean-up gang, he then declared that the sponsor had examined our salvaged products and was very happy with the results. Moreover, the company had found an outlet for the resale of the recovered merchandise. As a result, we stood to make *mucho dineros* for our hard work. He then sat down, yelled for the waiter, and ordered two more magnums of the Louis Roederer Cristal Brut.

Rico went on his long-awaited trip to Manila. Upon his return he told me the successful stories we had both envisioned. He said he was sure

155

he would be engaged soon and he already had the right wedding gown for his bride, who looked very attractive from the picture he proudly showed me.

I often reminisce about the Great American Salvage operation and secretly wish I could participate in another one. Mind you, I am not wishing for another Northridge earthquake.

Author's Note: *Regarding the term "structures," why don't we simply say "particles?" (Okay, maybe "fibers"?) I think lawyers and scientists have some kind of a conspiracy to use language that purposely keeps the ordinary public at bay. Being a scientist myself, I know that many of these technical concepts can be explained in terms that every Joe can comprehend. (I don't mean to offend anyone named Joe. In fact, some of the smartest guys I know are Joes.) But no, they must make it complicated and mysterious. I guess it makes them look even smarter and discourages ordinary folks from questioning their findings even when those findings are not on sound footing. I could go on and on, but this is also not funny.*

CHAPTER 16
BOSTON BLOODHOUND

Carl Sagan, the famous astronomer and author of *The Dragons of Eden* wrote on the subject of odor detection: "The most elaborate man-made device of this sort, the gas chromatograph/mass spectrometer, has in general neither the sensitivity nor the discriminative ability of the bloodhound, although substantial progress is being made in this technology."

How true, I thought. The facial protuberance of vertebrates, which we fondly call a nose, is a gift from the heavens. I always had the feeling that the Good Lord had blessed me with one of the finest, ultra-sensitive proboscises any human being can have. How could I take advantage of this olfactory gift from the Heavenly Father?

From stack sampling to odor sniffing

After several near-death experiences from climbing unguarded 150-foot-high metal ladders in frozen Upstate New York winters, I did a sanity check and decided that stack emission testing was not for me. I was going to be an odor expert.

Unlike stack testing, odor testing usually didn't require hard physical work. You did not need the stamina and the physical prowess of a steeple jack. I could do it. More important, it gave me the opportunity to use my nose, the part of my anatomy I could genuinely boast about in a crowded bar.

Soon I was running odor panels using such exotic machines as the "Dynamic Olfactometer," a device used to determine odor thresholds of stinky materials from a series of dilutions. No doubt my PhD in chemistry, combined with my ultra-sensitive proboscis, contributed to my rapid success in the field.

Olfactory prowess – a logarithmic function?
The difference in sensory perception between noses is probably logarithmic. In the air pollution business, mastering this concept could be worth more than your PhD or a PE (Professional Engineering license) or a CIH (Certified Industrial Hygienist certificate). You can make a good living just sniffing chemicals and sanitary odors all your life. I know of people who have retired from such activity and now live in beautiful beachside homes in the Bahamas and Florida.

So why not me? The lure of finer things and my penchant for chemistry convinced me that I could just smell my way through life. So it should come as no surprise that soon I was knee deep in odors: sniffing, measuring, and seeking correlations between chemical compositions and odor intensity (often there is none) and running odor panels, one of the most fun things in the whole environmental consulting business.

The renowned Bloodhound of Boston
ABC Environmental Services was a waste disposal company based in Baton Rouge, Louisiana. They were gearing up for a legal battle. A massive class action lawsuit had been filed against them that, if successful, could force the company out of business. ABC hired our company as air pollution and odor experts. My role was to assess the chemical odorants emitted from their incinerator stack and conduct comparative studies of all major odor sources in town and then rate them in order to identify the worst offenders. ABC also planned to put me on the stand as a technical expert for the fast-approaching legal hearings. The company VP said to me: "Man, you will be our designat-

ed hitter! Can you hit one out of the ballpark for us and convince everyone that this lawsuit is frivolous, targeting us unfairly?" He told me that "we will not deny that our plant emits odors, albeit for short periods of time during a malfunction. We will maintain, however, that ABC Environmental is not the only stinker in town. There are worse ones around. Why gang up on us?"

It sounded like a strange defense to me (admitting that you are the lesser offender among a bunch of bad actors), but I nodded in agreement anyway. After all, ABC was paying top dollar for my time and was putting me and my associate Pat Brogan up at one of the finest properties in town.

"I am pretty effective in such matters," I nodded to the VP in a manner uncharacteristic of me. As many of you know, I am a modest man.

"Jas, I know that, and that is why you are here, but can you tell me with one hand on the holy book, that you are the best nose, sorry best odor expert, in the whole United States of America?"

"Well, Mr. Jacobs, I don't think I can say that when you put it like that. I can confidently say that I am a qualified expert witness. But you know there are always ones who are smarter than me" (in this case meaning other people with bigger and better noses).

"Okay, then, do you happen to know one smarter than you who we can engage to assist you? This is big deal, man!" he emphasized.

I thought of all the super noses I had known or heard of in my life. In a flash I thought of a Dr. BB who was once described to me by a friend as the world's top odor expert. BB worked at a prestigious international consulting firm in Boston. Rumor had it that BB had performed some incredible odor projects including one that involved weeks and weeks of milling around Paris, smelling and cataloguing odors in the glamorous city renowned for its beauty, culture, and art.

"Go find Dr. BB, please!" the VP screamed.

An evening with the super nose

Arrangements were made for BB to fly to Baton Rouge on a Tuesday evening. Pat Brogan, who had just joined me at our firm in Detroit, and I were to pick up the good doctor at the airport, whisk him straight to a restaurant for dinner, and explain the mission.

Dr. BB turned out to be everything you would imagine in a renowned bloodhound—a tall six-foot frame, wide shoulders, a penetrating quizzical look, and the eagerness to proceed with the task at hand.

"May I suggest we check you in at the hotel, have a couple of drinks at the bar, briefly discuss the project during dinner, and then let you rest for the night because you had a long day? Tomorrow we will meet for breakfast, visit the ABC officials, discuss my odor sampling strategy, and review the hearing schedule." I laid it all out for BB.

"Yes. It all sounds great, but you are missing one thing," BB reminded me.

"What is that?" I was curious.

"You must first take me on an odor tour of Baton Rouge before we consume any ethanol or food so as not to compromise our olfactory senses."

Pat and I could not believe our ears. We looked at each other and without saying a word, repeated in silence, "An odor tour of Baton Rouge, Louisiana? Are you kidding me? How many people get to do this and get paid for it?"

"Great idea, Dr. BB," I muttered. "Let us get the show on the road."

I charted our route to cover as many odor sources as possible in the short time we had available.

I turned on the air-conditioning in the car to seek relief from the oppressively hot and humid Louisiana air.

"You must lower those windows," Dr. BB ordered. "We need to keep windows down as we are driving so we can detect odors when we are in the vicinity of an odor source, and when that happens, ease your foot off the pedal."

"Yes, Dr. BB," I responded in the manner of an obedient and trusted personal chauffeur.

BB started reciting his worldwide odor adventures, each one smellier than the previous. Pat, who was sitting in the back, leaned forward and listened in awe to BB's odor stories like when your grandfather used to tell you about all the strange and far-off lands he had traveled to during World War II when you as a kid had not even ventured beyond Kalamazoo (a town in Western Michigan), and even then only to visit Grammy and Grampy during the Thanksgiving holiday.

Suddenly BB cried out, "Stop! Stop right here!" His sudden outburst startled me to the point of skidding off the road.

"I smell styrene," he loudly declared.

Right on! Of course it was styrene. Even my inferior nose told me that. First of all, styrene, (also known as vinyl benzene, a basic raw material for the popular polystyrene), has a characteristically sweet odor. Second, it is a known ingredient in petrochemical emissions.

BB got out of the car, raised his head, pointed his nose toward the sky as if he were about to pray, and started inhaling deep whiffs of the thick odorous air surrounding the petrochemical complex. He seemed to be in a trance. His eyes closed as if in ecstasy, and why not? Styrene was not the only chemical there to savor. It was a chemical cocktail made in heaven by the angels themselves. There was also methyl styrene, polycyclic aromatic hydrocarbon rings with shapes just like those marvelous Lego® toys that so perfectly fit into each other. I sometimes wonder if the whole Lego® business was started by a mad organic chemist tinkering with the plastic models of the polynuclear rings?

To top it off, we sensed the possibility of the presence of hydroquinone, BTX, ethylene dichloride, esters, mercaptans, and aldehydes.

The time was getting late. I reminded BB that soon the restaurants would close. We needed to go back but would return the next day for one more odor patrol. BB agreed, but with one condition that we squeeze in one more odor source he had heard so much about before we go back. That was the city sewage treatment plant.

So we did. The sewage plant visit was every bit as exciting as we had envisioned. I was not very hungry for dinner that evening.

The next day we conducted another odor tour of the city and later said goodbye to Dr. BB with the promise to send him the odor data as soon as it was assembled. BB had a very early flight the next morning back to Boston, so he wanted to turn in early.

Sniffing bloated bags along a rural highway

Of the many odor panel opportunities I have had in my life, the Baton Rouge waste incinerator job was the most exciting. The fire-eating monster was promoted as a "state-of-the-art smokeless behemoth that will gobble up poisonous waste goo without belching."

The machine performed as promised most of the time, but every once in a while, especially in the evenings when families were sitting down for dinner, it would misbehave, and thick, black, acrid smoke would descend from the sky on unsuspecting citizens, obscuring their views and flavoring their barbeques with an unpleasant aroma.

Well, you can imagine the outcry! I sure can.

Being a physical-organic chemist, I can tell you that even a few molecules of skatole (the chemical mixture responsible for the disagreeable aroma associated with sanitary wastes) in the air will gag you. And me too. My proboscis gave me a sense that the air in some parts of Baton Rouge may have contained a hint of skatole.

People in the community, who apparently lost their sense of humor, retained a couple local lawyers and filed a class action suit against ABC Environmental Services for health concerns, loss of property value, loss of enjoyment of their homes, and, above all, impairment of conjugal happiness. This was bad. The company responded by hiring some hotshot Washington DC lawyers who had previously held important positions in the federal government. ABC also hired our company because we were technical experts in odor assessment and control.

Not allowed – alcohol, tobacco, chewing gum, perfume, and runny noses
As project manager, I marched into town with my colleague Pat, our suitcases bulging with air-sniffing gear and some impressive-looking devices, including "Dravniek's Dynamic Olfactometer," which was the mother of all sniffing devices! Rumor was that there were only a few of these devices in the country, and the granddaddy of all was in the possession of the Chicago professor who had invented it.

Pat and I would go out daily on a morning odor patrol around the city of Baton Rouge, stop at any source of stink we fancied, and fill a plastic bag (Mylar, it was called) with the foul air. Back at the decrepit roadside motel we rented, we would present our precious gifts of foul air to the waiting panelists. I had rented two motel rooms for this purpose. The first room housed the olfactometer, which was positioned in an ergonomically correct way in the middle of the room. An adjoining room housed the seven panelists, including the spare person—just in case we lost a panelist during our project. All panelists were regular Baton Rouge citizens lured through advertisements in the local newspaper with the promise of $9.50 per hour plus free donuts, sandwiches, cold beverages, and gas money, all just for sniffing bad odors during the daytime for up to two weeks. Overtime was rewarded with a 50% bonus.

On the appointed day the eager citizens arrived early and were treated with sparkling water (the aroma of coffee could compromise one's organoleptic senses) and an assortment of freshly baked pastries from a local Dunkin' Donuts® shop. Of the seven panelists, three were men and four were women. I made sure everyone had complied with the instructions I had sent earlier—no recent use of perfume, alcohol, cosmetics, spicy food, and definitely no runny noses.

Our panel looked fantastic!—professional, bathed, and well-dressed. One of the panelists was named Katrina, a shy and unassuming young lady. She was of French heritage (no surprise in the Louisiana Bayou country) but spoke no French. Katrina wore a dashing designer dress as if she were going on a hot date to a fancy night club. She was very attractive and soon became my favorite panelist, the reason being that she had the finest nose, a prime requirement to be a good odor panelist.

Pat and I set out to collect our air samples early Monday morning. Baton Rouge had no shortage of smelly places. Among others, we made sure to collect samples from the refinery, the city sewage treatment plant, and a sample downwind of the incinerator in question. Back at the cheap motel we rented, the magnificent seven waited anxiously. (Thanks to our client, Pat and I stayed at a much nicer hotel in the city.)

Protocol required injecting a measured amount of the odorous air into Draveniek's device and then diluting it manyfold with clean, filtered air. This created a known foul- to clean-air ratio. The diluted sample would then be delivered through a Pyrex cup that our odor panelists would wrap around their noses and take deep sniffs of, and if they smelled something noticeable, they would immediately jot down their response; fancier models let people press a button. They were also asked to describe the odor.

We started with the most dilute concoction, which even the most prized proboscis could not detect, and gradually increased the amount of foul air to the point where the panelists started noticing some odors. With every iteration, the mixture was rendered more disagreeable and, therefore, easier to perceive. This process determined what the experts called a "threshold." The threshold concept is very useful and is applied in the fragrance, cosmetics, and beverage industries.

Katrina had a fine proboscis

One of our last panelists of the day was Katrina. As required under the protocol, I presented her with the weakest (most diluted) odor concoction first, then increased the amount of foul air in the successive mixtures. Yet she detected no positive response (no odors). On sample no. 4, however, she suddenly perked up and shrieked, "I smell something."

"Can you describe it?" I asked.

"Oh, it is really hard to pinpoint," she said. "It is kind of earthy with a hint of apricot and peaches and freshly ground cinnamon," she said with the confidence and authority of a seasoned wine connoisseur.

"Can you tell me what it is?" Katrina inquired.

"Not right now because we still have two more to go, but I will reveal the source after we are done," I promised.

Katrina waited patiently, and as soon as the experiment was over, she reminded me.

"Okay, what is the big secret? Now that it is all over, can you tell me what it was you had me sniff? I don't think I have ever experienced this stuff before. Where did you collect your sample?" she inquired.

I hesitated and said, "You know, Katrina, as they say, sensory perception is all in your head. Some people like the smell of a certain thing while others hate the same aroma. By the way, did you ever see the 1992 American remake of the 1974 Italian movie called *Scent of a Woman*, starring Al Pacino?" I tried to change the subject.

"No, I did not see *Scent of a Woman*, but why are you trying to change the subject? So what is the big damn secret? Why can't you tell me?" She was getting impatient.

I could not dance around the issue any longer.

"Okay, Katrina, the sample you liked and described so elegantly was from the Baton Rouge sewage treatment plant, but if you liked it, it is okay with me." I continued, "The chemical responsible for the disagreeable odor is called skatole, a strong-smelling crystalline amine from human feces, produced by protein decomposition in the intestine and directly from tryptophan. When diluted sufficiently, the foul-smelling molecule actually becomes pleasant, and I am sure you have heard of sufficiently diluted skunk oil being used in expensive perfumes and, by the way, did you ever give your boyfriend the highly advertised French cologne called 'Jovan,' rumored to contain traces of pig sweat? It is claimed that no woman can resist a man wearing a little bit of pig sweat."

"Okay, okay, that is enough. I don't need a lecture in chemistry and I don't need some fancy chemical doctor telling me what I should give to my boyfriend for a present. You can save your scholarly advice for another day."

To say that Katrina was irritated would be an understatement. I tried to calm her down, but to no avail. The damage was done. She stormed out in a hurry. Her boyfriend was waiting outside. The car was running. Maybe they had a hot date planned.

The next day I was forced to use the alternate, who happened to be a forklift driver at a local petroleum refinery. I made sure that he did not carry the coker smells permeated deep into his thick epidermis. Luckily the man wore no odors, but alas, he did not have the fine proboscis that Katrina was blessed with.

Katrina never came back. I still miss her.

CHAPTER 17

EXPERT GAMES

Some of the best and the brightest scientists and engineers in the field of industrial hygiene are involved in what is known as "toxic litigation." Lawyers call these Toxic Torts, which simply means lawsuits brought by people who feel that they have been injured or their health or lifestyle has been impaired in some way due to exposure to toxic, noxious, or nuisance chemicals.

Toxic tort litigation has consumed a large number of the best minds in the IH profession at a time when their skills are most needed in controlling toxic exposures and protecting workers through training, monitoring, or using good industrial practices. You cannot blame them. The pay doing litigation support work is far better than the pay from chasing chemical poisons in factories. I do not deny the benefit of the involvement of many of my learned colleagues in litigation work because they interject sound science into an otherwise unscientific process, but their departure from our routine work is still a loss to the industrial hygiene profession.

Much misinformation is a part of the worker health and safety debate. During heated political fights, science and facts oftentimes take a back seat to sensationalism. What a pity! Many rules, regulations, and decisions about complex technical and scientific issues are not being made by the scientists, but by politicians, some of whom have probably never taken a course in chemistry, physics, or biology and certainly have not studied Bayesian statistics or Monte Carlo simulations.

One side benefit of this ever-expanding litigation is the injection of much-needed humor into our stressed-out society. Litigation in the environmental health and safety world—the only field I can talk about from personal experience—is replete with absurdities and could provide endless fodder for late-night comedians. For some reason, comedians have not caught on to this.

Expert testimony work is touchy, as opposing lawyers seem to have a mission to make the expert who has no business being in a courtroom look like an idiot. Of course there are exceptions. I have been cross-examined by opposing lawyers who were respectful, cordial, knowledgeable, and tactful. The good ones made me think, scratch my head for the right answer, and, every now and then, admit my shortcomings.

Although I know my stuff, some lawyers say I do not always make a great expert witness because I cannot keep my mouth shut. Anyone who has ever been a witness in a trial, regardless of whether as an "expert witness" or a "fact witness," has been advised not to offer information that has not been requested. The lawyers want you to just answer the question—not to expound or add details unless requested.

The term "fact witness" has always intrigued me because it implies that the other kind of witness, namely the "expert witness," does not use "facts." This implies that expert testimony is mostly high-quality BS. I don't think that is true, but I will let my readers be the judge. In spite of my reservations, I still relish being involved in toxic tort litigation, whenever there is an appropriate opportunity to do so. Humor and drama abound in courtrooms.

My interest in legal proceedings has been fueled by my daughter Monica who sends me quotes and anecdotes she has collected from recorded legal proceedings. Monica is not a lawyer, but between her bachelor's and her master's degree studies, she spent one year working as a paralegal at a prestigious environmental law firm in Los Angeles. My favorite ones include:

Lawyer: *So the date of conception [of the baby] was August 8th?*

Witness: *Yes.*

Lawyer: *And what were you doing at that time?*

On another occasion:

Lawyer: *Doctor, how many autopsies have you performed on dead people?*

Witness: *All my autopsies are performed on dead people.*

Toxic chamber

In the mid-1980s when I was working in Detroit, Michigan, I was called to testify in a case in the Upper Peninsula of Michigan that involved formaldehyde exposures. Usually I am hired by the defense (typically a manufacturing or chemical company), but I do not confine myself to being an exclusive defense witness because if you always work for the same side, you get tagged as a "hired gun." I detest that label.

If I see merit in a case and I feel qualified, I would consider taking a case for either side. In this particular case, I was the plaintiff's expert. The plaintiff's attorney—the attorney who also represented our client—had drilled in my head, "Dr. Singh, please comment only when you are asked. Refrain from offering information that is not asked of you. I know you have lots of facts in your head, but please keep the facts to yourself unless you are asked. Jury trials can become a game of wits. Anything and everything you say could be held against you." This was a tall order for me, but I said I would try to restrain myself.

The litigation in Traverse City, Michigan, was related to the release of toxic fumes from urea-formaldehyde insulation that had recently been installed in the home of an elderly couple. The insulation was injected into the walls of their home to help ward off cold during the bone-chilling Northern

Michigan winters. If installed properly, the chemical foam cures and hardens in a reasonably short amount of time. The two primary components of the chemical mixture embrace each other tightly to form a nontoxic bond. If the ratio of the two compounds is correct and the proper installation conditions are observed, no formaldehyde fumes are emitted from the cured resin.

If, however, the mixture is improperly prepared and incorrectly applied, the result can be a slow-acting toxic gas chamber from which the occupants cannot escape whether sleeping, eating, or playing cards with friends. Apparently the elderly homeowners had been living inside such a gas chamber for a while after the chemicals were pumped into their walls and roof. This had caused them many health difficulties, and they felt that their only recourse was to sue the installer who they contended had incorrectly pumped the foam into their home.

The defense team brought in a hotshot attorney from Washington, DC. The lawyer represented the institute whose member companies produced the chemical components of the insulation foam.

In my opinion, the installation job was botched. The hotshot lawyer from Washington, DC, unfortunately, had a different opinion. His position was that the chemicals in question were as harmless as the air we breathe and the water we drink.

Formaldehyde, he thought, was a gift from God. It was made of the same life-giving elements of oxygen, carbon, and hydrogen that the Lord gave us in abundance—the elements of our body that sustain all life.

The assertion that the elements of life could do no harm under any circumstances was such a load of crap that I was irritated to no end. My basic intelligence was being insulted! I was not going to let him get away with such statements. Despite my attorney's repeated warnings, I decided I would offer some facts not asked of me.

The hotshot attorney asked the judge to put me on the stand. Hotshot asked me many questions, including where and how I was born (implying that my birth did not occur in a proper medical facility and that the inexperienced midwife who delivered me to this world could have botched the procedure, causing hypoxia [oxygen deprivation], which might have resulted in brain damage, thus jeopardizing my ability to properly comprehend complex chemical concepts).

He questioned all of my qualifications, including my education at some of the most reputable US and Canadian institutions, my experience, and my certifications. I put up with all of the questioning like a nice obedient boy and did not say any words other than those necessary for the required response.

Hotshot then came up with a brilliant idea to destroy my credibility. He requested the judge's permission to allow me to write part of my answers to his questions on a blackboard in the courtroom. He argued that the rural jury needed to "understand the complex chemistry directly from Mr. Singh."

"Thank you, Mr. Singh, for agreeing to share your chemical knowledge with us," the attorney said. He adjusted his wide, multi-patterned necktie, and tugged at his three-piece suit. "Mr. Singh, do you know the chemical formula for formaldehyde?" he began.

Before I answered him, I felt compelled to correct him. "By the way, counselor, it is Dr. Singh." Hotshot took a small step back. The judge had a faint smile, which he quickly concealed.

"Yes, sir, I know the formula. Would you like me to write it on the blackboard?"

"Yes, would you please, Dr. Singh?" The hotshot sounded more respectful after being reminded of the doctor thing. I wrote the formula. "HCHO"

Hotshot lit up like a lamp. "Well, Dr. Singh," he continued. "Aren't those the basic elements of life [oxygen, carbon, and hydrogen] that you and I have in our bodies, given to all of us by the Good Lord? And much of our body weight is composed of those very life-sustaining elements, wouldn't you agree?"

"Yes, I would agree," I replied.

Hotshot jumped into the air, waved his hands, adjusted his ugly tie, again tugged at his brand new, three-piece suit, and exclaimed, "Well then, Dr. Singh, wouldn't you agree that these life-sustaining elements couldn't hurt anyone?" He proceeded to say that if they could hurt anyone, the Lord would not have given them to us, but he stopped in the middle of his thought. He thought better of it.

At this point I looked at the jury. The experts had told me that you always keep one eye on the jury to see how the citizens were reacting to the arguments. I could see that Hotshot was having an impact. Several of the jury members were nodding in agreement every time he uttered his baloney life-sustaining argument and especially when he invoked the Lord's name. I realized that damage was being done. Jurors were being impressed. I started thinking, *How do I debunk this crap?*

Perhaps I was taking too long. The judge turned to me and said, "Dr. Singh, please answer Mr. Hotshot's question." (I am certain the judge did not call him "Mr. Hotshot," but I do not remember what he did call him.)

"No, sir, I would not agree that the elements in question could not hurt anyone. The fact is that those same elements can kill and have killed thousands of people," I responded.

Having been in such trials before, I fully suspected Hotshot to object to my statement, but he did not. Instead, he challenged me, "How is that, Dr. Singh?"

Without answering, I stepped up to the black board and wrote on the blackboard: 'CO.'
I looked at the judge and the jury and continued, "These are carbon and oxygen, the two basic elements of life. The human body is largely composed of these two elements, yet hundreds of people all over the world die every year of exposure to CO, commonly known as carbon monoxide."

"Objection! Objection!" Hotshot shouted and leapt up from his chair.

"Sustained," the judge said emphatically and banged on the desk with his gavel.

He turned to me and said, "Dr. Singh, please answer only what is asked of you."

"He started it, your Honor," I said sheepishly.

Several of the jurors smiled, and I know two of them gave me the virtual and universal thumbs-up sign. I apologized to the judge.

Hotshot immediately said to the judge, "Your honor, I have no more questions for this witness."

"You may be excused, Dr. Singh. Thank you," the judge said. He gave instructions to the jury and sent them to deliberate.

My part was done. The next day I took the flight back to Detroit. One week later I found out that the defense had proposed a settlement that included fixing the problem for the plaintiffs. The couple would get to stay at a nearby hotel in the resort city of Mackinaw City, Michigan, while their house was being repaired. They would also be compensated for any mental anguish and discomfort they had suffered. I never knew the exact amount of the settlement. That was not important to me.

Cluster bombs

Later in my career I was called to work on another interesting case. A manufacturing company was producing a product using an antiquated technology. The age-old process generated a potent cocktail of chemicals. Some of those were alleged to be cancer causing to humans if inhaled or even contacted through bare skin. The exotic compounds had big impressive chemical structures that looked like Lego toys. The chemistry was complex and their health effects poorly understood.

The State of Maryland had cited the company for exposing their workers to polycyclic aromatic hydrocarbons. As I recall, the state regulatory agency had not measured anything, but based the citation mostly on visual observations and some anecdotal information. The citation against the company was under the "General Duty Clause." The General Duty Clause permitted the agency to cite an employer if, in the agency's opinion, there was a health and safety risk to the workers.

The company decided to fight the citation. They hired my company to assist them in quashing what they considered to be an unjustified citation. The company made an appeal on the grounds that their emissions did not violate the OSHA standards nor endanger any worker's health and/or safety. The company engaged one of the most successful law firms in Baltimore to defend them. I do not recall the legal details, but a hearing was scheduled in a district court to argue the case.

I flew to Baltimore to meet with the company's lawyers and one of their local technical experts. As usual, my lawyers grilled me on my knowledge of the issue and my opinions on the technical merits of the case. After a full day of debating, we felt we were ready for the hearing.

The hearing was in a conference room. Nearly twenty people were present, including the judge, several lawyers, three company officials, and Mr. Jensen, the director of environmental health and safety, a court recorder, and at least two others whose roles I don't recall. Before the issue of worker exposure to polycyclic aromatic hydrocarbons came up, the state's lawyer decided to express his concerns.

Henry was about sixty years old. Dressed in baggy, old-fashioned clothes, Henry reminded me of a country lawyer in a popular TV series, although Henry looked sloppier than the TV character. Henry may have lost a few pounds just prior to the hearing because he was having difficulty holding onto his trousers. He was constantly adjusting his pants.

Henry's strategy was to show the court that this was not the first time that the company had been a target of a state citation. His contention was that the company had a pattern of violations of the state's environmental regulations. Henry asked the judge if he could interrogate Mr. Jensen, the company's EHS director.

Mr. Jensen expected this. He had organized all his documents neatly on the table and was ready for the grilling. Henry looked straight at Mr. Jensen, pulled his trousers up, which by now had slid down a full two inches, and fired his first volley.

"Isn't it true, Mr. Jensen, that not too long ago your company was cited for polluting the waterways of the State of Maryland? May I remind you, sir, that your company was discharging improperly treated wastewater into the navigable waters of the state? The discharged water contained certain organic compounds in a concentration that was in excess of the regulations. In addition, your discharges contained some carcinogenic poly . . . compounds," Henry said haltingly. He was having difficulty pronouncing the formidable chemical names while trying to read his crumpled notes.

"Yes, sir, we did have slightly elevated levels of solids, oil, and grease concentrations, which we promptly corrected by tweaking our wastewater treatment plant process. We are in compliance now, and, by the way, sir, we never had any carcinogens, I mean cancer-causing chemicals, in our discharge. That was wrongly alleged," Mr. Jensen said in a measured tone.

Henry was not satisfied with the sincerity of Mr. Jensen's answer. He probed further. He again adjusted his pants and asked, "Mr. Jensen, the state never received any evidence of those changes or any proof that the violations had been corrected. Would you please tell the court what you did that makes you sure that the treatment plant is now working properly?"

Mr. Jensen launched into a detailed explanation of the workings of the treatment system. He explained the primary treatment process and then the state-of-the-art secondary treatment process, which he said was recently upgraded to halfway tertiary status.

For lawyer Henry's benefit, Jensen decided to explain in more detail the workings of the secondary biological oxidation process. In his refined Scandinavian accent and his patient manner, Jensen explained, "In the biological treatment process, you inoculate or 'culture' your organic wastes. After a while the tiny microorganisms go to work. Each individual microbe is microscopic and may not look like much, but collectively they are like cluster bombs. Their effect is amazing, I mean in a nice way. The little microorganisms break down the big, organic molecules and render them into innocuous compounds such as carbon dioxide and water vapor."

Before Mr. Jensen could finish, Henry gave another tug to his slipping trousers and jumped into the discussion. He was getting impatient with all this micro talk. He said to Jensen, "Mr. Jensen, I am no engineer or chemist. I am a simple lawyer (no kidding!). I don't understand how these 'microorgasms' work, but I do know that as of now we have no proof of compliance from your company."

Henry's last half of the sentence was totally lost in the roar of laughter from the people in the courtroom. Even the judge could not contain himself. It was a riot! The balding grandfatherly judge adjusted his glasses, still laughing, turned toward Henry, and said, "Henry, I know what is on your mind this morning, but this early?"

Another roar of laughter went around the courtroom. Henry was bewildered. He had no clue what had just happened. He approached the judge and whispered so only a few of us sitting in the front row could hear. He said to the judge, "Clive, what the hell just happened here?"

The judge got closer to him and whispered in Henry's ear low enough that no one could hear him.

"Oh sh!t!" Henry could not keep his voice down enough. He

clutched his head between his hands and said to the judge again in clearly audible tones, "Clive, please call a brief recess."

Without hesitation, the judge complied. One thing was obvious. Henry and the judge had a good relationship.

"Ladies and gentlemen, this court is adjourned for the day. Tomorrow, we will reconvene at nine o'clock a.m., sharp," said the judge. "Bang!" went the gavel.

The next day, when I went to the court house at nine o'clock a.m., I found out that the two parties were in a conference in Judge Clive's chamber. We were asked to wait. At ten o'clock a.m. everyone emerged from the judge's chamber. The group included Mr. Jensen and Henry. Judge Clive picked up a piece of paper, looked at the audience, and said, "Ladies and Gentlemen, we had a conference. The two parties have met and discussed the case further. The state withdraws its case against the company. This case is settled." Another bang of the gavel and we were dismissed.

I never presented my testimony after preparing for several days. My fee was smaller than it would have been had the hearing gone for several more days as anticipated. I am not complaining.

Six months later I was having lunch with Lisa, a senior environmental manager at the company where I worked, in a well-known eatery in San Francisco. Lisa had invited an important local client to join us. The issue was an upcoming lawsuit against the company that was somewhat similar to the case in Maryland. We were to provide expert assistance. Lisa related my Henry story to our guest. The well-mannered and composed client went crazy. He was laughing so hard that people around us were staring at us.

While still laughing, he kept saying, "Cluster bombs! I imagine they are tiny, but imagine the damage they can do if you have the right partner." Lisa, who is not easily embarrassed and is probably more liberal and tolerant than any professional woman I have ever worked with, was blushing and embarrassed. She was helpless. This was a big client. She just kept smiling. Luckily, the waiter brought the check. Lisa paid the bill and we said goodbye to the client who was still mumbling (although in barely audible tones) "micro-orgasms" and "cluster bombs!"

Two weeks later Lisa wrote me saying, "Jas, the Cluster Bomb man told me to say hello to you. He did not remember your name, but asked me to bring the funny guy along next time we meet."

"It will be my pleasure," I wrote back to Lisa.

CHAPTER 18

ACRONYMITIS

On Sunday, February 10, 2013, I went downstairs to get the Sunday newspaper. The headline in the bulky weekend edition of the *Honolulu Star Register* read, "PLDC Must Go."

As I was deciding which story to read first, I wondered what PLDC was. The front page main headline had to be about an important issue, but what in the world was PLDC? I had never heard of the acronym before. If I read further, I was sure I would find out what kind of animal this PLDC was. Oh well, I thought, I am a news junkie. All I watch on television is CNN, *60 Minutes*, late-night comedy shows, and news specials. I would need further research to figure out the meaning of PLDC (which, by the way, I never did).

The use of acronyms is spreading in our society like a virus. We are being bombarded with acronyms that we may have never heard of before and we may never hear of again in our lifetime. Mastering such abbreviations adds nothing to our knowledge, our comprehension, or our value to society. The use of acronyms is mind numbing. The supposedly user-friendly computer industry has greatly added to the height of this electronic Tower of Babel.

The newest strain of this "acronym-virus" epidemic is emanating largely from text messaging and is not confined to younger people. It has inflicted well-educated and articulate grownups. I know many senior colleagues who have been touched by this virus. Younger individuals are abandoning written English en masse and replacing the beautiful, rich language with little messages composed mostly of single roman letters, characters, dots, and few, if any, words. There seems to no longer be a need for real words. I worry that one day perhaps whole books will be written in acronyms, lines, dots, and funny little cartoon faces called "emoticons." I must admit that some of my friends and younger relatives do make an effort to improve their single syllable sentences by embellishing them with emoticons.

According to Wikipedia, an emoticon is a pictorial representation of a facial expression using punctuation marks, numbers, and letters to express a person's feelings or mood. Emoticons are often used to alert the recipient of a message to the mood or temper of a statement, and can change and improve the interpretation of plain text. (No kidding!) Without emoticons, most text messages would be difficult to interpret because there are few words, phrases, or proper punctuation marks.

While I am thankful for the assistance of emoticons in making sense of the text messages I receive, I do not mean to imply that I adore them. Initially I thought they were cute. But not anymore. Without words expressing a thought, emoticons sometimes look stupid and insulting. Why didn't the sender just say "I like you" instead of pasting a weird-looking symbol?

And sometimes in their haste, people slap on a wrong emoticon. On one occasion I received a text message with a sad face attached to the greeting that was supposed to convey Happy Birthday. I thought the sender should have sent a smile because it was my birthday and life had been good to me.

Or was this supposed to be a sad day? The sad expression of the emoticon could have been intended to remind me that I was not getting any younger and the sender was sad for me.

I decided to disregard the emoticon and celebrate my birthday anyway. I sent a text reply to my well wisher, "Thank U. UR grrr8. CU soon." She did not write back.

Emoticons galore

I am usually not desirous of attaching mood statements to my messages. Most of my electronic mail is business related and is a simple statement of facts, whether the facts are happy or not. Once in a while when someone has wasted my time by sending me an acronym garnished with mood symbols, I am not amused. In such situations, I look for an emoticon that says "screw you," although with a smile. I have not found such an emoticon. If ever invented, however, it could be the most popular symbol in today's hectic world.

I can understand the temptation of using acronyms and emoticons when texting. You could be short of time. Sending a letter could be too costly, you might be spelling challenged, or you may have a limited vocabulary. Still such shortcuts are not excusable.

Excessive and unnecessary use of acronyms and emoticons compelled me to look for reasons why texting addicts inflict repetitive stresses on their wrists and thumbs and some annoyance to the recipients of their missives. After some soul searching I identified the following causative factors:

- Laziness
- Inability to spell
- Inability to compose
- Low regard for the recipient
- No value for other people's time
- Attempting to fill a void by using an emoticon
- A need to convey the impression that they are busy
- Trying to look funny, but being devoid of a true sense of humor
- Using emoticons to express emotions because the sender has none

The spread of the acronym virus concerned me so much that one day I mentioned it to a physician friend. "This acronym thing is spreading like a disease," I said. "The prospect of being hit every day with new acronyms frightens me. Many acronyms make me feel incompetent and behind the times. I feel I may be losing it. I have actually found myself sitting in some meetings where a super nerd was throwing around strange acronyms as if they were household words. I thought everyone understood those except me. So I sat there like a dummy. If I asked what this word was, my ignorance would be exposed. Others would think I was behind the times, a relic of the past," I continued.

At one of the meetings one brave soul raised his hand and asked an acronym-happy nerd, "Excuse me, but what does this acronym stand for?"

The acronym-happy guy gave the acronym-ignorant man a disdainful look and said, "Sorry, I thought everyone in the room knew the word" (it was not a word; it was an acronym) and then spelled it out.

I thanked the man for posing the question and plunged into the discussion realizing that this was a subject that I knew. I just did not know the acronym.

I turned to my physician friend and said, "You medical people need to come up with a name for this malady (the impulsive and excessive use of acronyms.), develop diagnostic tools, and find ways to remedy the affliction."

"We do have a name for this disease," he replied. "It is called 'Acronymitis.' We also have a medical term for the associated disease that you just described. That is 'acronym phobia,' which is the fear of acronyms. The causative factor for this disease is also known. Acronyms make you feel incompetent and inadequate even though you may know the subject. Unfortunately, the virus is still gaining strength with no known cure in sight. People seem less and less willing to make an effort to spell out things. The phenomenon is already out of control."

Acronym phobia sounded like a genuine medical term. I wondered if there was an emoticon available to convey the acronym phobia mood. None was listed.

This discovery of acronymitis has opened a whole new line of terminology to me. This terminology could prove useful to anyone in interpreting and dealing with acronym-virus affected communications. I have already benefited from the "acro" terms I coined, which include the following:

Acro-messaging (short text messages composed mostly of acronyms)
Acro-emotions (emotions expressed through emoticons without any human involvement)
Acro–writing (same as acro-messaging but not in the texting format)
Acro-love letters (amorous communications without using real words or expressing real feelings; the benefit is that you won't get stung by real rejection)
Acro-hate mail (abusive communications without being abusive; an acro-ignorant person will not feel the pain)
Acro-decoding (this is simply an acronym dictionary that is voluminous and growing)

"V love U and V C U soon"

I received the above text message from a favorite niece in India who is one of the smartest members of my extended family. Her parent's high hopes for her future are justified. She has good manners, earned excellent grades in college, has a sweet demeanor, and respects her elders. These attributes were drilled into her head by her loving but strict parents. I showed the text message from her to one of my young colleagues at work. She said, "Oh, how cute. Your family in India must love you very much. How nice that they will be coming to see you soon in Hawaii!"

I was startled. I had failed to interpret the important message encoded in the symbols. Where was it said that they would be seeing us soon in Hawaii? I had learned English, but now I needed to learn acro-writing. I was not willing to learn it. I wanted my niece to learn English instead.

After receiving the "V love U and V will CU soon," I started thinking about how lovable "V" must be. She is in love with "U" who must be a special fellow.

I contacted my niece through a text message and expressed my irritation at the acro-messaging. She replied "that is why V love U uncle. U R very funny. Can't wait until V C U and auntie."

Technical disciplines such as computer sciences, engineering, and health and environmental sciences have also contributed mightily to the spread of acronymitis. Acronymitis is making life miserable for millions who do not use the abbreviated jargon. Every time they receive one-syllable communication, they run to Google to find out what it means. Even Google cannot keep up with the onslaught. Furthermore, time wasted in searching for someone's favorite acronym contributes nothing to the economy and the betterment of society.

Every day I experience a barrage of messages. Being a senior professional, I am on the email distribution list of numerous technical groups, associations, professional communities, marketing teams, and promotional subgroups. I feel privileged to be included, but with this honor and privilege come hundreds if not thousands of email messages that have little relevance to me and about which I have nothing to contribute. Half of the time I do not understand the content.

A few weeks ago I received one such message. The subject of the message was never stated in words, only by five letters of the alphabet. Although the matter was supposed to be of critical importance to my profession, I had no clue what was meant. The coded email had been sent to about one hundred people who speedily replied. A dozen people engaged in the critical discussion. Almost everybody hit the "reply all" button with their responses, swelling electronic traffic by the logarithmic scale. Finally someone begged for mercy and asked the meaning of the acronym. The sender should have spelled the acronym to begin with.

The PCB breakfast

Polychlorinated biphenyls (PCBs) were a big environmental health and safety issue a few years ago. The foul-smelling dark, oily fluids were used extensively in electrical equipment and caused environmental and worker health concerns. Organizations rushed to understand and control the toxic substance; technical bulletins, informative memos, and warning messages flooded the mail.

My company decided to conduct a series of informational seminars about PCBs in several major US cities as a public service. More importantly, we wanted to promote our services related to the testing and remediation of PCB contamination. I was usually the featured speaker at these half-day seminars.

One seminar we presented was at the Hilton Hotel in downtown Atlanta, Georgia. Our intent was to invite a large number of people from companies on our client list. We would give the attendees a free breakfast and a well-rehearsed, fact-filled technical briefing.

At the end of the seminar, the attendees would receive the company's promotional materials and our business cards. The Atlanta session went as planned except for a couple minor incidents.

The first surprise was the big sign in the Hilton Hotel lobby that read "ABC's PCB Breakfast" and the room number. As usual, Bob, the president of the company and my boss, welcomed the guests, introducing himself and the company. He then introduced the topic and the speaker (me) and motioned for me to come to the podium. I thanked him and greeted the guests who numbered around sixty. I asked if each guest would take a couple of seconds to introduce himself or herself, tell the audience what brought them to this seminar, and name what they hoped to get in return for the investment of their time.

"Our company had a recent transformer leak, resulting in the release of PCBs. I want to understand what the risk is and what we should do," a man responded. He had summed up the purpose of the event in two short sentences. I nodded in agreement. That was the precise reason for the event.

Another person said: "We do not have a problem, but we have some old transformers and capacitors. Anything can happen any day. Old capacitors could get overheated and burst. We are trying to understand everything about this issue. Your seminar is timely, and thank you for the wonderful spread (breakfast)."

"This is very generous," another man responded graciously and then said, "However, I may be in the wrong place. I am in the printed circuit board (PCB) business. I have never come across an informational seminar on printed circuit boards with a free breakfast included. This is nice, but I just realized that I don't belong here. Do you want me to leave?" the man asked.

"Oh no, sir. You are most welcome to stay. You are our guest," I said. The man's name tag identified him as being from a big, well-known company in the Atlanta area that had a tremendous potential for future business. He stayed and learned nothing about his brand of PCBs.

I was almost at the end of the introductions when two strange-looking characters walked into the room. They had very long hair, disheveled clothing, and a bewildered demeanor. As they looked at me and the audience, they had a "deer in the headlights" sort of appearance.

Then they noticed the buffet table and immediately proceeded to load up their plates with food—reconstituted scrambled eggs, sausage gravy with biscuits, and fresh fruit. Afterward they calmly settled down into chairs to eat their breakfasts.

"Good morning, gentlemen, and welcome to ABC's PCB breakfast. Would you gentlemen share with the audience what you expect to get out of this seminar?" I began. I was suspicious of their motives but did not want to challenge their presence. At a distance they appeared to be bums and not legitimate business professionals. But I wanted to be cautious. As we have learned, looks can be deceiving! At an earlier seminar I had a couple people dressed like poor college students who were later introduced to me as directors of environment, health and safety for major corporations. I had learned to be cautious.

"Awesome, man" the older and scruffier one replied. "What we wanted most of all, you have already delivered (the sumptuous breakfast buffet). I mean, this is one hell of a spread. We haven't had this kind of feed in years! Second, we saw your 'PCP breakfast sign,' and we are interested in this angel dust stuff. We have been snorting for a while. Everyone is telling us we could croak any day, but I don't believe that. Look at Gary and me. We are healthier than most of the other dudes in this group."

PCP, an arylcyclohexylamine derivative (⌬), is a dangerous and illicit drug with strong hallucinogenic effects. On the street it is called "angel dust."

"We figured you might be handing out free samples" he finished.

The audience was shocked! There was some nervous laughter followed by stunned silence. Most of the audience then figured out that these guys weren't your typical EHS professionals.

I told the fellows, "Please enjoy your breakfast, but there should be no talking and no smoking." The two winked at each other at the word 'smoking.' They behaved as well-mannered school kids during the rest of the presentation. At the end they came to the dais, shook my hand vigorously, said, "Awesome, man," and left.

This occurred twenty years ago and was at a time when acronyms were easily sorted out.

Acronymitis has spread a million-fold since then. Now it is out of control.

CHAPTER 19

SEWAGE CHEMISTRY FOR DUMMIES

Awhile back, someone wrote a story or a book titled: *"How to Discuss Sewage with Women."*

The title struck me as odd, discriminatory, and plain weird. Why would it be unique? Is sewage only relevant to men? Are women immune to this?

Sewage is an extremely important issue in a modern society. Just think of the mounting sewage disposal problems with growing populations, megacities, and the impact on the economy. Sanitary waste disposal today is probably a multi-trillion dollar industry employing hundreds of thousands of our brightest men and women.

So why a book with such a title?

The more I thought about this, the less I was surprised. Our society seems loathe to discuss subjects that we consider unpleasant no matter how relevant or critical to our daily living and our civilization as a whole.

The tendency to ignore topics we consider unpleasant extends to odors, chemistries of decay, and human anatomy in general. This is even manifest in the clumsy debate on sex education in schools and the awkward efforts parents make to avoid mentioning the correct anatomical terms to children presumably because some of the body parts we were born with are strange and mysterious to the opposite gender. As a result, we adults come up with weird acronyms (sometimes funny and even cute, I must admit).

I will leave sex education to others. I am the sewage expert. However, I cannot help thinking that I could simplify and render palatable sex education for even the most conservative citizens by devising a numbering system for the various organs and the sexual postures all the way from the simplest and not so offensive, to the very advanced and disgusting orientations. Just imagine: *Kama Sutra by the Numbers*. Perhaps that could be my next book: *Sex Education for Dummies*. It would be X-rated. To do that I would have to revisit Khajuraho: the ancient Indian city in the State of Madhya Pradesh and the location of the ultimate examples of erotic sculptures (see photo). For now, I am content with my G-rated stories.

A bath in the park. *Really?*

In our clumsy efforts to avoid using words like urination and defecation, we have coined the terms: number one for urination and number two for defecation. I know of children who did not know the correct terms for these routine body functions until they were in high school. Of course when we are adults, we can easily avoid such terms by just being excused to go to a washroom or a bathroom as most Americans would say and never have to delineate any reason at all. It is usually not relevant to others. However, when you are a kid, the distinction can be important to parents.

Using terms such as "going to the bathroom" can be confusing to non-Americans. I remember one time when my wife and I were with a German postdoctoral fellow who was a researcher at the National Research Council of Canada in Ottawa, Ontario. We were visiting the beautiful Gatineau Park in Hull, across the bridge in Quebec. I told my German friend that I wanted to go to the bathroom before we headed back. He seemed utterly surprised.

"You want to take a shower? Right now here in the park?" He seemed incredulous.

"No, Jurgens, not a shower. Only number one, and it won't take long."

"This sounds even more puzzling to me," Jurgens said. He was disappointed because he took such pride in feeling that he was getting pretty good with his North American English, including colloquial and casual chit chat, but he had no idea what we were talking about. Apparently the little pocket dictionary he was carrying with him did not explain number one.

Upon explanation he thanked me. He was delighted to learn useful American slang that you could only learn when you were in America. Or in this case, in Canada.

The numbers game

Linda and I worked together for a large insurance/loss prevention company. After that we both moved in different directions. I was working for a large environmental company in California while Linda took a position with the US Navy as a safety expert. We were good friends. The bonding force in our relationship was our common weird sense of humor and irreverence for most things others considered important. Anyway, it was always refreshing to see Linda and shoot the breeze, as they say. I savored every opportunity to do this.

On one of my international trips, I was returning back to Los Angeles when I missed my next connection. This meant several hours of delay, resulting in a long layover at the San Francisco airport.

I immediately thought of calling Linda, who lived not far from the airport, to suggest we meet for dinner somewhere near the airport.

"Sounds great, Jas, but wow! What a short notice. Not sure I can do that."

"Why not?" I was being pushy as usual. I dreaded spending four hours in the airport after twelve hours in an airplane already.

She said, "You know, I would love to see you, but I don't have a babysitter. Had I known earlier, I would have arranged for one."

Linda had a five-year-old son she once brought to the office. He was delightful and well behaved and I thought that now a couple of years later, he would be acceptable company during dinner.

"Bring Mark with you. He is a fun kid."

"Are you sure? He can be a pain in the hind quarters sometimes."

"It will be fine. This will be a quick informal dinner," I assured her.

We decided to meet at the nearby International House of Pancakes, apparently a favorite eatery with both Mark and his mom.

Traversing through the International House of Pancakes (now called IHOP) menu required speed and decision making. There were scores of entrées with only slight variations, some differing only in the sauce offered and whether you wanted the banana slices on the top or cooked into the dough. The process was so repetitive and confusing that for efficiency, the restaurant devised an elaborate scheme to order your food simply by calling the number you fancied. There was no need to mention the food item.

My favorite from the extensive list was number two, which I believe was the Hawaiian special with banana pancakes sprinkled with powdered sugar and macadamia nuts and the gooey white banana syrup on top or offered on the side.

The waiter came around to take the order. We were ready. Linda ordered her usual favorite. Mark wanted the "Kiddie Special." When the waiter came to me, I was ready.

"Number two with coffee," I said, perhaps a bit too loudly.

Mark almost fell out of his seat.

"Mom, Mom, he said number two," he shrieked.

"It may be his favorite, Mark. I am sure he had it before."

Mark shrieked even louder.

"He wants number two; yikes!"

Several heads turned in our direction.

An elderly white-haired man with a grandfatherly look was sitting next to us. Apparently he overheard all of this, turned to Mark, and said.

"Mark, I think your dad is right. You will like it too. It has bananas, macadamia nuts, and nice sweet banana syrup."

"He is not my dad. He is mom's friend," Mark strongly protested. He was getting angry. He thought the adults were poking fun at him. His dad would never say anything so disgusting.

By now Linda understood the reason for the kid's distress. She pulled Mark close to her and whispered something in his ear, and I heard her say to Mark, "Now you say sorry to Uncle Jas."

"Sorry," Mark muttered without looking at me. He was not willing to accord me the uncle status yet (he later did). "I thought what you said was the 'bathroom word.'" Mark seemed to apologize then, although he did not say that.

Our food arrived. Entrée number two was as good as always.

We had a nice discussion. Linda had changed jobs and was now working for Cal-OSHA, the State of California agency responsible for enforcing the Federal OSHA Health and Safety Standard in the state. She seemed to like it.

Toward the end of the dinner, Mark leaned toward his mother and said, "Mom, I need to go to the bathroom."

Linda pulled him close and asked in a hushed voice, "Number one or number two?"

Mark seemed embarrassed. "Number one, Mom!" He seemed irritated.

"Yes, she sure is a number-one mom," I completed his sentence.

He smiled and said, "You are funny." It was time to head back to the airport.

Manure experts

On the way back to the San Francisco airport, I could not get the number one and number two discussion out of my mind. The very next day I was scheduled to give a deposition in San Diego related to a lawsuit alleging "malodors" (very unpleasant smells) from a newly constructed sewage treatment plant on the outskirts of the beautiful seaside city.

The next morning I had to explain to lawyers the chemistry of sanitary odors. What were the offending chemicals released from the new plant that made the neighboring affluent community on the hill so upset? Was it just the foul odors, or was there a more serious issue of exposure to toxic chemicals that could cause irreversible health problems?

I started thinking about nitrogen and sulfur, the two elements critical to sustain life, yet a source of so much annoyance, avoidance, and litigation. Of the malodors that can send property values crashing and people abandoning their homes and communities, sanitary wastes were usually the worst offenders.

Because of the abundance of such legal claims and the resulting financial impacts of such litigation, at one time the study of offensive odors became an obsession of mine. That obsession earned me the unwanted title of "sh!t expert" from some of my closest friends. At least I think that they were my friends. This was not flattering, especially when being introduced to some of the female delegates at the conferences I attended. My favorite line at all those embarrassing moments was: "Somebody has to do it." This response usually worked and did not seem to jeopardize my chances of getting to know them better.

This well-deserved reputation, however, had landed me the interesting expert testimony case I was headed for in beautiful San Diego, California.

Note taking or doodling?

The new sewage treatment plant on the outskirts of San Diego was billed as state-of-the- art—no odors, no smoke, and no excessive noise. Expert landscaping made it look like another suburban building perhaps housing a library named after a dead politician or city father.

The problem was that the planners and designers failed to take into account the new million-dollar homes up on the hill a stone's throw downwind from the plant. The technical geniuses also forgot that the malfunctioning of the equipment could shower the occupants of the expensive homes with a cocktail of highly odorous mercaptans, indoles, skatoles, and other aromatic compounds. Had someone performed detailed air dispersion modeling using some of the sophisticated computer models available, the results would have been predictable.

On certain evenings, the affluent folks up on the hill would be sitting in their backyards ready to barbeque and open a vintage California chardonnay or pinot noir. Suddenly they would be showered with indole, skatole, and dimethyl sulphide cluster bombs from the malfunctioning treatment plant. This tended to spoil the evening.

I don't want to imply that these stink bombs would cause serious health effects. In reality, the health effects from exposure to such foul-smelling molecules are often negligible because odor thresholds for many chemicals are an order of magnitude lower than the chemical's toxic threshold. This means that the bad odor could make you gag at concentrations too low to cause deleterious health effects. Nevertheless, can you imagine smelling rotten eggs and worse when you are sipping a vintage *fumé blanc*?

One of the house owners in the most direct line of fire filed a lawsuit against the owners of the treatment plant. Her attorneys sought expert professional help. I went to the plant with a pen and a writing pad. I looked at the design, operating procedures, the performance, and the available test data and talked to the plant supervisor and the engineer. They were very cooperative and professional. I came home satisfied but noted that I had not written down a single word in my note pad.

Not to worry. That is how I was most of the time, I said to myself. In the evening after dinner, I would summon my vast memory bank and recreate notes. But this time I was not very successful. I was interrupted by several phone calls. I did fill up five pages, but it was all doodling, mostly the hexagonal polycyclic aromatic benzene rings, the Lego-like playful structures that I always draw to release tension.

Testimony was set for the following Tuesday. Despite my lack of written notes, I was ready.

A couple of days before my testimony, my trusted but overly cautious office manager Audrey showed up in my Cypress, California, office, pointed out a piece of paper, and said:

"Jas, this is a court order for us to submit all the documents, reports, and even the handwritten notes to the lawyers on the San Diego stink job you are doing and remember, you cannot take out anything from the files, no matter how unrelated." Audrey pronounced this with the authority of a federal marshal.

"But Audrey, you know that I do not have any reports, not even any handwritten notes from my visit to the plant. All I have is five pages of doodles and the single word 'BOZO' written in the middle of the doodles on one page."

"Then you must submit to the court your five pages of doodles and the Bozo, and please Jas, no sorting out, no picking and choosing, and no purging. You are one of the smartest men I have ever worked with and I don't want to see you in any legal trouble. May I remind you what happened to Richard Nixon for purging eighteen minutes of the Watergate tapes?"

"I remember, I remember. Okay. I am not going to jail," I assured Audrey. "You send the file, as you are holding it in your hand. I will not even look, and if any embarrassing things are in there, so be it. I will deal with the consequences of those lawyers dragging me in the mud and destroying my reputation built over thirty years." I was trying to humor Audrey, who seemed stressed out, which was a normal phenomenon for her.

"It is not funny at all," Audrey thundered. "You should not be saying such things even in jest." Audrey, who was fifteen years older than me, gave me the needed tongue lashing.

"Okay, okay, Audrey. Go ahead and send the file via FedEx," I told her.

(It is funny. When you know someone is competent, sincere, and loyal, nothing they say will offend you. That was my relationship with Audrey.)

At the trial the opposing attorney, as expected, asked me to identify the file he was holding in his hand. He came straight to the point: "Dr. Singh, are these all the reports and the notes you have on this case?"

"Yes, sir, this is all."

"But Dr. Singh, all I see here is some interesting doodles, and in the middle is the word 'Bozo' circled twice. Does the word Bozo relate in any way to any individual in this case?"

"No, sir, not at all. Bozo is my favorite character and essentially a stress reducer, and doodling is my way of concentrating during an important and stressful phone call. I do not recall what prompted me to write down the word Bozo in this case, and I might add that to me, Bozo is not something derogatory. I like Bozo the clown. He is cute and relaxing and reduces tension whenever I think of the word."

"Okay, Mr. Singh, understood." The judge seemed a little irked with my Bozo sermon, and I know from experience that a good witness never offers extra information not asked of him or her.

The defense attorney was not about to quit. He kept hammering: "Dr. Singh, wouldn't you find it unusual to go to the site but take no notes and not conduct any tests of any kind at all?" He was challenging my methodology and my thoroughness.

At this point, I felt that a little display of my technical prowess was necessary. I turned toward the judge while keeping the lawyer in my line of sight and retorted, "No, sir, I did not think any of that was necessary because I don't need to spend $10 thousand to $20 thousand dollars of my client's money performing gas chromatography and mass spectrometry using a variety of detectors like a thermal conductivity and electron capture detector just to prove what we already know. Sh!t smells bad!"

I had to apply quick brakes and rephrased: "Human waste smells bad; this is a well-known fact."

"Okay, Dr. Singh," both the judge and the lawyer said in unison.

The lawyer immediately said to the judge: "Your honor, the witness can be excused."

Another Win! I congratulated myself silently while still on the stand.

Watch your language

To complete this chemistry lesson, it is important to know that the two usual suspects most responsible for much of the stink and misery are indole and skatole. Yes, I could mention the hydrogen sulphide (most people know as the rotten eggs smell) and its more aggressive cousin called mercaptan, but in terms of shear unpleasantness, none can match the potency of the winner: SKATOLE

Without going into an advanced organic chemistry lecture, every person should know what the chemical beast looks like.

So, Ladies and Gentlemen and Boys and Girls, I present to you the stinker:

This is SKATOLE:

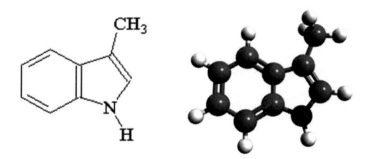

Note: Skatole has attained such prominence in our society that it has become a favorite expression in literature to describe foul vocabulary and gutter language. To say that someone has a foul mouth, we simply say he or she has a "scatological" vocabulary.

This completes Sewage Chemistry Class 101.

CHAPTER 20

ISO-CERTIFIED BROTHEL

Credibility is important to every trade. The environment, health, and safety field is no exception. The public wants to be assured that a company is conducting their business in an environmentally safe manner and protecting their workers from health and safety hazards.

The International Standards Organization (ISO) 9000 Quality Management Standard was an important step in assuring the quality of manufactured products. ISO 9000 was first published in 1987. The ISO 14000 certifications, an offshoot of ISO 9000, were a milestone in the environmental standard–setting process. These standards were developed to help organizations design management systems to prevent adverse changes to air, water, or land as a result of their operations and to comply with the applicable laws and regulations.

While the ISO 14000 standard brought much attention to environmental management, including the protection of biological species such as mud turtles, it did little to promote worker health and safety. Don't get me wrong. I love mud turtles (and I love mud). It is a shame, but to some people, humans rank lower in the food chain than animals as far as preservation of the species is concerned.

As recently as 2010, the International Standards Organization said there was no prospect of an ISO-type international standard to secure worker health and safety.

Efforts have continued, however, toward voluntary and lesser health and safety protocols. Several countries have banded together to agree on a "non-global" system called the OHSAS 18000 series. The 18000 systems have become quite popular in some Asian countries. My own observations of the OHSAS certification process have not given me a warm and fuzzy feeling. I have not seen a correlation between the OHSAS certified facilities (and for that matter the ISO-certified facilities) and a (hoped for) higher level of environmental health and safety protection. One reason could be that much of the resources expended to date have been spent on show and tell, public relations, hoopla, and chest thumping.

Take me to the City of the Brave Lady

Korat is the second largest city in Thailand. Korat is also the name of the largest province in Thailand, more commonly known as Nakhon Ratchasima. The city covers an area of 25,000 square kilometers. There is a lot to explore around Korat in addition to the stunning natural beauty and historical destinations. Korat's historical and architectural significance is evident from the hordes of German tourists in the area. I believe that next to the Thai language, German may be the most common language spoken inside the Korat hotels. Hotels are what I know most about in Thailand.

The symbol of Korat is the Thao Suranari, or Lady Mo, who is also known as Ya Mo ("Grandma Mo"). Lady Mo was the wife of the deputy governor of Nakhon Ratchasima. Suranari is an ancient Sanskrit word that means "brave woman." True to her name, Thao Suranari is known to have protected the city and the kingdom of Siam against the invasion of the Laotian Army. Female descendants of Lady Ya Mo still exert power and influence in Korat.

Note to American readers: "Lady Ya Mo" of Korat should not be confused with "Yah Mo B There," the song written by American singers James Ingram and Michael McDonald and producer Quincy Jones. "Yah Mo B There" was released as a single in 1983 and peaked on several UK musical charts. More recently, the song was referred to in the 2005 film: *The 40-Year-Old Virgin*.

Getting to Korat

The reason I have travelled to Korat so many times is that Korat is now a significant industrial center. This is a big change for an area which only recently was mainly engaged in the cultivation of rice, tapioca, peanuts, corn, jute, sugar cane, and sesame.

Over the past twelve years I have been to Korat a dozen times to provide health and safety consulting for several US multinational companies. It is not an easy destination to reach despite being the second largest city in Thailand. The only practical way to get there is to hire a taxi in Bangkok and endure the long, bumpy, yet scenic, four-hour taxi ride to the city.

If you search Google and Wikipedia, don't expect to find much useful information about Korat. You will be inundated with "Korat Cats," which are popular in the west but not useful information for a business trip. Also, your Korat search will bring up the word Nakhon Ratchasima, which is another name for Korat but also the name of the province in which the city is located. Sort of like saying "New York, New York" (which educated readers will know is a massive city in a large state). To avoid taking an unplanned trip through half of Thailand, do not say "Nakhon Ratchasima" to the taxi driver. Tell the driver to take you to the City of Korat.

Karaoke on the 5th floor

On my second trip to Korat I asked Somchai, my host at the electronics company, if he could recommend a hotel in town. Somchai told me that Korat had no fancy five-star hotels like the ones in Bangkok, but the four hotels he mentioned were the best that Korat had to offer. All four hotels boasted four stars. From my past experience, the number of stars in an Asian hotel is not necessarily a relevant ranking of quality. I inquired further. I wanted Somchai to state his preference.

"The four hotels I mentioned to you are essentially equal, but the hotel on the hill is ISO certified and the costs are the same," he said.

"Well, that settles it. If the ISO-certified hotel costs the same as the uncertified ones, why would anyone want to stay at an uncertified property?" I reasoned.

As soon as I cleared customs in Bangkok, I hired a taxi for the trek to Korat. After the four-hour ride, I arrived at the ISO-certified hotel on the hill. It was late. I was tired and hungry after the long plane trip from the USA and the four-hour taxi ride. I checked in, dropped my bag in my room, and went down to the lobby to get a drink and a snack.

The lobby bar looked dead. It was past eleven p.m. I asked the hotel receptionist, whose name tag read "Amporn," if there was another restaurant or a bar in the hotel. Amporn spoke excellent English. In the Thai language, her name translated: "Girl from the sky." It was a fitting name for her.

"There is a beautiful bar on the fifth floor. It is a karaoke bar and you can get all kinds of drinks. The hostesses are gracious. Some of them may even be college educated and they talk good English. They entertain foreigners all the time, including many Americans. I noticed sir, from your registration, that you are from Hawaii. That's a lovely place."

I have known about the karaoke (カ ラ オ ケ) bars in Japan where karaoke is an interactive entertainment in which the amateur singers growl along with recorded music using a microphone and a speaker. What they sing is typically a well-known pop song minus the lead vocal. I have been to only one real karaoke bar in Japan, but I have been to several replicas in Malaysia and Thailand. Unlike the Japanese establishments, singing plays a minor role or no role at all in the Malaysian and Thai versions. Often times these places are a cover for more intimate games. So when Amporn directed me to the karaoke bar on the fifth floor, I was not sure what kind of a sing-along would be there. Nevertheless, I wanted to go. I was thirsty for a Singha (Thai) beer.

I thanked Amporn for the suggestion to go to the fifth floor. "Sounds like that's what I need after the grueling trip," I said.

"I am sure you will like the karaoke bar and, by the way, sir, it is certified. The whole hotel is what they call 'ISO certified,' including the restaurants and the bars. I am not sure what it is all about, but I am told it is a big deal, as you Americans would say. It is all about quality."

"Yes, it is about quality and it is a good program. This is the reason I selected your hotel."

"You will not be disappointed," Amporn assured me.

I stepped out of the elevator and entered the fifth floor. On one side of the floor was the dimly lit bar with a dozen men and women huddled around the bar laughing and yelling. Most were Thais. Two of them looked European. On the opposite side of the bar was a small, well-illuminated swimming pool. A young boy was skimming the leaves that had just landed on the water. I was not sure why he was cleaning the pool at this late hour. Perhaps they had occasions when an inebriated customer needed a dip before going home.

Two well-dressed young women and a middle-aged lady were sitting around the large marble table by the poolside, chatting away in Thai. All three women were strikingly beautiful. The middle-aged lady, who was dressed in jet-black silk, got up and greeted me with that most graceful of the gestures, the Thai greeting that requires properly folded hands and a respectful and ergonomically correct 30 degree bow. "Hello young man. What can we do for you?"

I instantly liked the place. It had been awhile since a well-dressed attractive lady had called me "young man." Amporn was right. This was the place I needed to visit. The lady in black appeared to be in her mid-forties and wore heavy jewelry as if she planned to go out on a hot date. I decided she was the Madam, the Mama San.

"I would like a beer and some snacks. I know it is late. Maybe some potato chips or peanuts would do."

Mamma San motioned to the bartender inside, and without any verbal communication, she conveyed to the bartender the correct snack and beverage order. Then she turned to me and said, "Well, young man, while we are waiting for your drink and the snacks, may I present to you some of our loveliest hostesses. You can decide which one you want to have as your dinner partner and then perhaps take her to your room to chat." Mamma San winked at me (I had no idea why she did that). "All my girls are charming with excellent manners. I am sure Amporn told you that the place is certified," she said proudly.

"What is this certification all about?" I asked Mamma San, although I knew it must be the ISO certification that the hotel had been advertising all over town, including the thirty-five-foot long banner plastered all over the building claiming it to be the only "ISO-certified hotel in Nakhon Ratchasima."

"Oh, sir" (perhaps because of the potential she sensed in me, she switched to "sir" instead of "young man," which I had preferred). "I am not sure what the certification is all about, sir, but it had something to do with quality and quality is what we offer. I suppose you can judge that for yourself."

Before I could open my mouth, about a dozen well-dressed and attractive young ladies appeared out of nowhere. They formed a semi-circle around the marble table that supported a big, still-unfurled umbrella. Mamma San's ladies greeted me with cupped hands and big smiles while standing in the graceful postures perfected after repeated drills.

It was dramatic and overwhelming. I stood up and returned their greeting with folded hands and the 15 degree bow I had learned while travelling in Japan. I did not say a word. I just froze. It was surreal. I was in a dreamland. It reminded me of one of my favorite Bollywood movies when I was growing up in India. The scene was where the actor Dev Anand, the famous heartthrob in India at the time, was at a college campus in Delhi or Bombay.

The heartthrob was admiring a bevy of gorgeous women, young and flirty and dressed in colorful saris. Suddenly, as is the custom in Indian films, Dev Anand erupts into a playful song, which goes like this:

They all are beautiful. They are all young.
Who do I lose my heart to?
They all reside in my heart;
Which one should I call?

As a young boy, I was a movie addict and I would daydream of such encounters. Only a handful of men in Bollywood could ever experience this in real life. Here I was, finally, experiencing the exact thing as Dev Anand. It was not a village kid's daydream. It was happening to me. I was surrounded by a dozen of the prettiest women in Thailand. They were every bit as charming and gorgeous as Dev Anand's beauties forty years ago. This was real. My mother had always told me: "*Jas, if you study hard, you will realize your dreams.*" Mom was right.

"Okay, young man, which one is for you?" Mamma San was trying to pull me out of my trance. I was taking too long. The ladies were frozen in the graceful but ergonomically awkward static posture.

I was also frozen in my not-so-graceful posture and rendered immobile by the surreal phenomenon surrounding me. I was still in a trance and was thinking of the words by Firdausi, the famous Iranian poet, and uttered by the Indian Mughal Emperor Shahjehan upon his visit to the Shalimar Gardens in Kashmir.

Gar Firdaus Ba Ru-e-Zameen Ast... Hami ast O, Hami ast O, Hami ast.
(If there is a heaven on earth, it is here, it is here, it is here.)

Mamma San's voice brought me back. "I know it is not easy. Who do you pick? They are all beautiful. It is not easy to decide which lady touches your heart." Mamma San was repeating Dev Anand's lines almost verbatim as if she could read my mind. It was spooky.

The ladies were getting irritated. They were losing their grace under ergonomic stress.

I woke up and realized that after all that, this too, was just a dream. I was not flirting with Bombay coeds. I was in a Thai brothel where every man with a wallet was a heartthrob.

I said thanks to the pretty ladies and turned to Mamma San.

"Yes, it is impossible to pick and it is almost midnight. I need to go to work early tomorrow. Tomorrow night I will come back under better circumstances."

Mamma San had heard such promises before. She had been around. She said in a song like lyrical manner, "They all say that, but tomorrow never comes. Do try to come back tomorrow when you are more relaxed."

"I will," I assured Mamma San, knowing full well that tomorrow would not come.

Disappointment was all around. I had wasted their time. I decided to give Mamma San a nice tip sufficient to buy each of the ladies a drink. Mamma San thanked me, but drinks were not what the beauties had hoped for.

The Synergist Magazine story

When I returned to the United States, I decided to write a story on my strange experience for the *Synergist Magazine,* a monthly publication of the American Industrial Hygiene Association.

The certification process was being exploited and manipulated. More money was being spent on show than substance. Liquor stores and cigarette venders were claiming to be ISO 14000 certified in an effort to sell more booze and tobacco. You might ask: Why not an ISO-certified brothel?

ISO and OHSAS certifications are third-party audits. This means that a consultant can get certified to certify others. Once certified, the certifier has free reign to grant the coveted piece of paper as they see fit. During my performance review of scores of health and safety surveys of operations in several Asian countries, my checklist always included the question whether the company I was visiting was ISO certified or OHSAS certified. It seemed like an important status to achieve.

On one occasion I was told: "No, we are not OHSAS certified, but if you come back next month, we should be." I went back after two months. The plant was now certified. They bought expensive full-page advertisements in the local paper. They distributed little OHSAS buttons to all employees. They congratulated each other. They gave a half day off to some of the employees.

I carried on my health and safety survey of the plant as commissioned. The plant had many health and safety hazards. Among them: unguarded machinery, long and frayed electrical extension cables transporting the deadly 220-volt energy that upon a touch would embrace you in a death lock. The lachrymator fumes of the heated glue made me cry. I gasped for breath from the acrid fumes emanating from the nearby unventilated oven.

Consolation was that the facility was OHSAS certified. The charts tracing the management systems were colorful and professional. The management system chart was so complex that it made me feel trapped and incompetent. Arrows were going in all directions. I felt like I had entered into a maze. How could I get out of the maze, and if I did, would there be a prize?

The chart had lots of boxes, large and small. Each box had somebody's name. A good number of the people in the designated boxes had left the company a long time ago. Two of the employees whose names and titles were still on the EHS organization chart were confirmed dead. Yet this was the chart that earned the company the coveted certificate and triggered the wonderful celebration.

I did not see many other documents, reports, or studies related to worker exposures, ventilation surveys, or engineering controls even though the need for such actions was clear. Nowhere in their well-arranged EHS files was there any mention of benzene, formaldehyde, asbestos, or the chlorinated solvents that permeated the workplace. Machine guarding did not surface as an issue even though two of their employees had lost fingers from unguarded machinery a year earlier.

I wanted to tell my story about the ineffectiveness of the OHSAS certifications. The fifth-floor ISO-certified karaoke bar was my window of opportunity.

As expected, my little article in the *Synergist Magazine* brought many responses. Most respondents told me it was funny but had a serious message. Two of the respondents did not like my story but for different reasons. One of them said that I was trivializing an important process. I was not. I was trivializing the exploitation of the process, not the process.

The second individual resented my presentation of the brothel example because it was a "lurid" subject in a health and safety publication. He said it was not funny. Discussing such topics could affect morals of younger people entering the field. I assured him that I was bringing up an issue relevant to our profession to the attention of adults. In support of my argument I mailed to him the guidelines from the New Zealand Department of Occupational Safety and Health Service (OSH) on "Ergonomics for Workers in the Sex Industry." The "Overuse Disorders" section of the New Zealand document proscribes steps to limit the risk of injury through workplace and equipment modifications or through administrative controls, strategies familiar to every professional industrial hygienist. I told him that he could download the complete document from the department's website: www.osh.dol.govt.nz.

The majority of people I heard from wanted to know the exact whereabouts and address of the ISO-certified hotel on the hill with the fifth-floor karaoke bar. I guess they did not believe my story and wanted to check it out for themselves. I did not oblige them.

Where have all the Apsaras gone?
The following year I went back to Korat. It was time for my yearly review of the electronics plant. As soon as I checked in at the ISO-certified hotel, I decided to have a drink. Against my better judgment, I headed for the fifth floor. I wanted to see if anything had changed. When the elevator door opened

and I stepped onto the fifth floor, my heart sank. It was dark except for one dim light in the corner where an old man was sweeping the floor and adjusting exercise machines. The machines had been installed in a feeble attempt to convert the karaoke bar into an exercise room. Gone were the ladies and the Mamma San. It was a sad place. The man introduced himself to me as "Amara," which in the Thai language means "immortal" or a "sage who had achieved nirvana."

"What happened here?" I asked Amara.

Amara shook his head, breathed a deep sigh, and said, "They are gone. The Apsaras are all gone." Then he looked up and asked me, "Do you know what Apsaras are?"

"Yes, indeed, sir, I know all about Apsaras. I have been all over Asia where Apsaras reigned supreme, to the point of being worshiped," I told Amara. "The word 'Apsara' probably comes from the ancient Hindu classic called the Rig Veda. Hindu scriptures point to the existence of Apsaras who act as the handmaidens of Indra and the dancers at his celestial court. It is rumored that Apsaras were sent to the earth by Indra to seduce ascetics who were becoming more powerful than the gods and had to be restrained."

Amara's comparing the bar ladies to the Apsaras made great sense. I had been searching for an appropriate term to describe the beauties that had surrounded my table the summer before. They were just like Apsaras. I never knew their names, but I know they would have exotic names like those of the real Apsaras, probably something like Menaka, Menakshi, Misrakeshi, Rambha, Purvachitti, Varuthini, Sahajanya, Chitrasena, Chitralekha, and Uravashi. Their eyes were like lotus leaves. Apsaras had slim waists and large gyrating hips that would perform intricate convoluted dances. Shaking their bosoms and casting their glances at the spectators, Apsaras would easily steal the hearts and the minds of their admirers.

Amara was listening intently to me as I explained all of these details. He also had deep knowledge of the Khmer history and culture. He had roots in Cambodia. He was enjoying my descriptions and motioned for me to continue. I told him that I became interested in the Apsara culture during my travels to Angkor Watt where I observed a few live performances. At one of the hotels was an exhibit of Cambodian art. A painting of an Apsara on display was so beautiful that I bought it on the spot. It now hangs in my dining room and always reminds me that only God could create such grace.

Amara interrupted me and said, "Then you know what I am talking about. The ladies on this floor were the reincarnation of the Apsara painting you bought in Cambodia. They are gone. Apsaras are gone for forever," he sighed again.

Amara did not want to stop talking. He had found someone to talk to after all those lonely evenings. He continued, "No one comes up here anymore except a lone exercise jock. They said this place did not fit with the main business of the hotel. It was undesirable. It was unsafe. I do not understand. The Apsaras were peaceful. They never bothered anybody. They never stole anyone's money. Apsaras don't hurt people. They bring goodwill and happiness. With all the criminals, thieves, and thugs around, why would they pick on the Apsaras?"

"I agree with you. Our priorities seem mixed up. It is easier to pick on the weak instead of the powerful. So tell me, what are your days like now?" I wanted to change the subject.

It was as if he had been waiting for this question all along. He delved into his life story. He was retired from an international company where he had worked for forty-five years until his retirement. Soon after retirement, he started working at the fifth-floor karaoke bar when it first opened. He liked

it a lot. The ladies were nice to him. They started calling him grandpa.

I started worrying that maybe it was my article in the *Synergist Magazine* that had brought the place down. Was I responsible for the Apsaras losing their livelihoods? I did not mean to do that. I am not a moralist. I feel no ill will against people who have different moral attitudes as long as they do not hurt others or threaten me.

I tried to console myself. The closing of the place should have nothing to do with my Synergist story. After all, how many people in Thailand would read the *Synergist Magazine* published monthly by the American Industrial Hygiene Association?

I do not accept responsibility for the flight of the Apsaras.

CHAPTER 21

BEHIND THE IRON CURTAIN –
INTRODUCTION

A Historic Visit to Stalinist Russia

Foundations of Stalinist Russia were beginning to crumble by the mid-1980s. Gorbachev's Glasnost Perestroika, meaning "openness," had started the process of cutting Stalin's metal chains that kept millions of Russians as prisoners in their own land. On December 26, 1991, the Union of Soviet Socialist Republics (USSR) was finally dissolved, acknowledging the independence of the fifteen Republics of the USSR.

In 1984, one year *before* Mr. Gorbachev started his political liberalization process, a delegation from the USA was assembled under the leadership of Mr. Chain Robbins, head of health and safety for US Steel. The purpose of the visit to the Soviet Union was to exchange technical views and information on worker health and safety with our Soviet counterparts. I was fortunate enough to be part of that delegation. It was a truly historic visit behind the iron curtain to discuss worker health and safety issues, which were not a priority in Eastern Europe in those days.

Our trip was led by INTOURIST, the official Soviet tourist agency that had the concession required to take foreigners around the USSR. Started in 1929 during Joseph Stalin's reign, the travel agency was suspected to be a tool of the KGB. Our official tour guides were two pleasant ladies who were fluent in Russian and English. One could never ascertain whether our chief tour guide and hostess, Svetlana, was a KGB agent, but she was as strict as you would expect from a KGB loyalist. With time, however, the human spirit of friendliness and understanding triumphed over political dogma and cultural prejudices and the US delegation parted with fond memories of our interaction with our Soviet friends, including our INTOURIST guides.

Several of the stories in the "Behind the Iron Curtain" series chronicle our interactions with many wonderful people, especially Svetlana, who mellowed remarkably over time.

Memories and notes, including my interview with CNN after we returned from our visit, lay buried in my files along with fading photo albums until recently when interest in the history of the industrial hygiene profession (or as the Europeans call it—"occupational hygiene") perked up. Industrial or occupational hygiene are the same thing; both are all about caring for the health and safety of workers.

217

In 2012, I dug up my old notes and photos and decided to share some stories about life in Stalinist Russia and the status of occupational hygiene in the Soviet Union in the nineteen eighties. The 1984 technical exchange trip to the USSR forms the basis for the following six stories. I encourage you to read these stories in the order in which they are presented. They are:

Behind the Iron Curtain - Part I
Alana does not live here anymore
Attempt to deliver gifts to a young Jewish woman in a Moscow apartment complex

Behind the Iron Curtain - Part II
Mission accomplished in Azerbaijan
Finding a recipient for gifts in Baku

Behind the Iron Curtain - Part III
Stolichnaya Express to Georgia
A seven-hour bus ride to Tbilisi

Behind the Iron Curtain - Part IV
Exchanging gifts can be hazardous
Ordering morning coffee through listening devices in a Kiev, Ukraine hotel

Behind the Iron Curtain - Part V
Big fat Armenian wedding
Crashing a party in Yerevan

Behind the Iron Curtain - Part VI
Romancing the stone
Our last day in Russia. Saying farewell to Svetlana, KGB friends, and other INTOURIST guides

CHAPTER 22

ALANA DOES NOT LIVE HERE ANYMORE

Behind the Iron Curtain – I

Rumors began to spread at the company in Detroit, Michigan, where I worked nearly thirty years ago—Jas is going to Russia as part of a US industrial hygiene professional exchange delegation!

I was flooded with requests and advice. "Be careful." "Do not stray from your prescribed path, as we want you to come back." "Take lots of pictures and bring home some Russian souvenirs, especially those cute wooden dolls." Someone requested a replica model of the Kremlin. And another person wanted a used Soviet military uniform.

The most intriguing request came from one of our employees who had recently joined the company. Her name was Deanna, and her family had recently immigrated to the USA from Russia. Deanna cornered me in the office one day and in hushed tones said, "Jas, will you do me a favor when you are in Russia?"

"Well, of course, Deanna. What can I do for you?"

"One of my best friends lives in Moscow. Her name is Alana. She is Jewish, just as I am. I haven't heard from Alana in a while. She used to contact me regularly, but I have heard nothing since last March. I am really worried. I wonder what has happened to her. I am unable to get any news about her or her husband Joseph. You know, anything could have happened to my friends. Things are not so good these days for Jewish people in the Soviet Union. I would like to give her some gifts to cheer her up."

The idea of looking for Alana sounded risky and spooky, but I thought Deanna knew the situation and she would not knowingly put me in harm's way. I believed she cared about me.

"Okay, Deanna, give me the things you want me to deliver to Alana. Pack them securely and I will deliver the package to her in Moscow," I assured her.

"Oh, thank you, Jas. You are a true gentleman!" *And a fool!* I completed the sentence under my breath, although Deanna did not hear me.

The next day Deanna tucked a carefully folder paper into my pocket, reminding me to keep it in a safe place, buried deep in the middle of other papers inside my briefcase. On the paper were the address and directions to Alana and Joseph's flat in Central Moscow.

As far as I know, this visit to Russia was one of the earliest international exchanges involving industrial hygiene and occupational safety professionals from the USA. In 1984 the Soviet Union was a closed country. It was still Stalinist Russia. Gorbachev's glasnost was just an idea at the time.

At the time of our visit, I believe the tourist agency INOURIST was still firmly under the control of the KGB. No one in our group could ascertain whether our chief tour guide and hostess, Svetlana, was an undercover agent, but she was as strict as you would expect from a KGB operative. Many of our experiences in Russia involved interactions between our group members and Svetlana. Happily, her authoritative conduct mellowed over time.

Our three-week tour of the Soviet Union started and ended in Moscow. We visited several governmental agencies and a couple academic institutions. Considerable sightseeing was featured in the tour and included several major Moscow landmarks, including the Kremlin, Saint Peter's Basilica, Lenin's tomb, and the famous superstore GUM. We were scheduled to spend only four days in Moscow.

The package from Deanna was 'burning a hole' in my briefcase. Finally, on the fourth day of our stay in Moscow, I decided that I must do something about this surreptitious package. The dangerous cargo had to be delivered before we left Moscow to begin our tour of other cities. Several of my new friends in the delegation advised me that I should not go on this adventurous (some said reckless) mission after dark, and if I was foolish enough to go, I should be accompanied by another person. None of my close confidants volunteered to go. Only Joe offered to accompany me on this mission.

Joe was a brash, outgoing, and outspoken safety professional from Union Township, New Jersey. Eager and unafraid, Joe was very excited to accompany me on this clandestine mission. Before leaving America, our delegation had met in a hotel near the JFK airport in New York for a let's-get-acquainted dinner. It was there that I first met Joe. He made his mark quickly with his aggressive but friendly demeanor. After dinner we all voted Joe as the "delegate most likely to get pepper sprayed" while in Russia. Actually, he came close several times. His irreverence also quickly earned him the nickname "Toxic Joe," only partly due to his professional career. We were not surprised to learn that his employer made some of the deadliest pesticides and insecticides known to mankind.

Toxic Joe and I quickly became buddies. I admired his raw humor and daring, but above all, his irreverence for just about everything. At the same time I worried about his penchant for getting into trouble, sometimes bordering on recklessness. Nevertheless, we got along great.

On the evening before we were scheduled to leave Moscow for Azerbaijan, I informed Toxic Joe that I was in possession of surreptitious cargo. I showed him the address scribbled on the carefully folded and already crumpled piece of paper Deanna had given me back in Michigan. At this point I had not seen the contents inside the carefully packaged bundle, although Deanna had told me it was cosmetics, undergarments, and some costume jewelry. I explained to Joe that the mission was to find Alana, discretely pass her the goods, and give her a big hug from her best friend back in the USA.

The prospect of hugging a total stranger in Russia really inspired Joe. I think he also sort of fancied himself as the Peter Graves figure from the *Mission Impossible* TV show, but without the high-tech gizmos. I was seriously hoping that our adventure wouldn't end in failure, death, or capture. That would seriously impact my employer's TRCR statistics.

"It is awfully nice of you to do this for someone you don't even know. I didn't think you had this in you," he said to me.

I felt mildly insulted. "Why do you say that? I am a compassionate person, always feeling the pain of others."

"A pain you are," he reminded me.

After ten p.m. that night, Joe and I set out to find Alana.

Moscow's streets, always dimly lit to conserve energy, seemed dimmer and more foreboding than usual. Perhaps this paranoid feeling was partly due to the risky mission we were about to undertake. The taxi driver we had hired at our hotel was not very helpful. He vaguely knew the address. It took us forty-five minutes to drive from the hotel to Alana's apartment complex.

Somehow he found the place, but he dropped us off outside the big steel gates that led into a rectangular apartment complex. He could have driven us inside the gate but he chose not to. Instead, he said he would wait outside, and he warned us not to take more than fifteen minutes on whatever mischief we were undertaking. If we did not come back quickly, he would not be there. He could not take the risk of being spotted by the police or the KGB. This was a Jewish area and was watched closely. Some of the tenants may have been Jewish activists. The taxi driver demanded half the fare before we ventured inside the gate. We paid him and promised him a big reward, including a bottle of Stolichnaya Vodka if he did not abandon us on the dark and scary Moscow streets.

We entered the grim complex. Not a soul was outside any of the apartments. No bikes. No swing sets. No evidence of play or the joy of children. The atmosphere was eerie. From the address, we figured that Alana's apartment (number 6) was on the second floor. After successfully negotiating the dark staircase leading to the second floor, we started reading apartment numbers one after the other. Someone in apartment number 5 partially opened their blinds, scanned our silhouettes, and immediately closed her shutters. We were getting anxious as we realized we were almost at Alana's place. We found apartment number 6 and froze!

A giant padlock was on the doorknob. The nameplate on the door had been hastily covered over. Even spookier was the black shellac brushed sloppily on the giant lock. The hardened material that by now had seeped into the keyhole would easily advertise if someone had attempted to tamper with the oversized lock.

To make sure we were looking at the correct apartment, we examined the nameplate one more time. The Roman letter "A" was still visible on the not-so-carefully covered plate. We figured the "A" stood for Alana.

Suddenly, the person in apartment number 5 parted her curtains again and opened the window just a few inches. A wiry old lady whispered in passable English, "I don't know who you are, but I know that you are not supposed to be here. *Alana does not live here anymore!* I don't know where she went. I know you are foreigners, probably Americans." She had probably noticed our blue jeans. "If you love your families, get out of here as soon as possible."

The old lady abruptly closed her blinds and window.

Joe and I looked at each other and then scanned a 360° panorama. We imagined people standing in shadows with trench coats and hats but saw no one. In haste we descended the dark stairway, being careful not to make any sound, and tiptoed our way out of the gray, haunting prison-like apartment complex.

To our great relief, the taxi driver was still waiting but anxious enough to have his engine running. He motioned for us to get into the already slow-moving vehicle.

The return ride to our hotel did not take nearly as long as the outbound trip because Moscow's traffic had ebbed. We reached the hotel and thanked the driver. Joe placed some rubles into the man's hand. Not-so-toxic Joe then reached deeper into his pocket and squeezed a few US dollars into the man's palm, this time for the promised Stoli Vodka. The Russian smiled and said a pleasant, "Dosvedanya."

In the hotel bar, a few hardcore members of our group were chugging shots of Georgian Pepper Vodka and being loud and obnoxious, keeping up the tradition of wild cowboys from America. They were glad to see us back with all our body parts still intact but seemed shocked to find that we were still carrying our surreptitious package.

My worry now was what to do with Deanna's gifts for Alana. Until now I had felt no strong desire to peek into the package, but now I wanted to know what the merchandise was that I had been carrying with me all this time. I did not want to take the goods back to Deanna in Michigan, but to whom should I give them? Knowing a little bit about Deanna's tastes and her affection for her friend Alana, I figured they were probably expensive feminine items, probably the best products that her new country, America, had to offer.

Joe and I decided to go to my room to inspect the contraband. As we expected, they were hard-to-get items in Russia at that time (L'Oreal cosmetics, a gold bracelet for a petite woman's wrist, and several neatly folded Victoria's Secret undergarments with bright pink labels). The undergarments piqued Joe's interest. I reminded him that this was not the proper time for his close inspection.

I was still scared and very concerned about Deanna's friend and her husband. The plan was for me to call Deanna back home after the drop and convey the news in some kind of a code. There was no doubt in my mind that any call to the USA would be listened to by our hosts. The subject of my call would be sensitive and potentially very damaging. I decided that I would not make the call. I needed to talk with Deanna in person when I got home.

Toxic Joe and I agreed that we must get rid of the contraband in a nice way. After considering our options and doing a risk assessment, we decided that at our next stop we would carefully select a suitable woman, ethnicity notwithstanding, who would be the recipient of some fine American merchandise. We were willing to risk the possibility of getting into a little bit of trouble while executing this noble task.

We found a worthy recipient during our next stop – Baku, Azerbaijan.

CHAPTER 23

MISSION ACCOMPLISHED
IN AZERBAIJAN

Behind the Iron Curtain – II

Baku, Azerbaijan, the oil-rich Caspian city—the second stop on our tour—was worlds apart from gray and gloomy Moscow. People in Baku had smiles on their faces, a phenomenon rare in Moscow. Even the women seemed friendlier in Azerbaijan than in Moscow.

Baku is the largest city on the Caspian Sea and in the Caucasus region. Over the past thirty years, since our delegation visited the city, Baku's urban population has quickly expanded to more than two million inhabitants. In 2000 the inner city of Baku, along with the Shirvanshah's Palace and the Maiden Tower, were named as UNESCO World Heritage sites.

Baku is the scientific and industrial center of Azerbaijan. Many Azerbaijani organizations are headquartered in Baku. More recently, Baku claims to have become a world-famous entertainment center. In 2012, the city hosted the 57th Eurovision Song Contest. Baku will host the European Games in 2015. According to some travel publications, the city is now among the world's top-ten destinations for urban nightlife.

The Fire Temple

One of the highlights of our sightseeing in Baku was a visit to the amazing Fire Temple at Suraxany, also known as Ateshgah. Known mostly as the "Parsi Fire Temple," the site contains multicultural and multi-religious inscriptions and artifacts. The inscriptions are mostly in Sanskrit or Punjabi, and at least one is Persian. The inscriptions contain the dates written in several ethnic calendars, including the Sanskrit Samvat (संवत) calendar1802 (१८०२) and the Arabic Hijri 1158 (١١٥٨) calendar. Scholars ascertain that the various inscribed dates correspond to the period from 1668 CE to 1816 CE. Some scholars postulate that the structure may have been built by Baku Hindu traders because the inscriptions in the temple written in Sanskrit (in Devanagari script) and Punjabi (in Gurmukhi script) identify the site as a place of Hindu and Sikh worship. The Punjabi inscriptions are quotations from the Sikh holy book, the Adi Granth. The Sanskrit inscriptions are known to be from the Sat Sri Ganesaya Namah.

Farah, the non-KGB tour guide

Russian women have been much maligned in the west, most of it, no doubt, by Russian comedians in the USA in attempts to garner cheap laughs. Throughout my adult life I have heard unflattering comments about their stout stature. The famous comedian Smirnoff used to say: "Russian women are like trucks," and while the audience waited patiently for the punch line, he would say: "That is it. That is the joke."

Such comments no doubt arise from the fact that Russian women have excelled at sports such as weight lifting, discus throw, javelin throw, shot put, and hammer throw.

However, when we arrived in Russia, we found that such characterizations were unfounded. We met many attractive women during our trip. Both our Moscow guides, Svetlana and Skaya, were attractive. Svetlana could even have been described as pretty if she would have just shed her titanium socialist armor and been just Svetlana (as we discovered later).

The ugly myth was shattered once and for all when we arrived in Azerbaijan. Our local guide in Baku was Farah. Svetlana was still with us. The tour protocol required that we should always have two guides on our itinerary, our "national guide" (Svetlana) plus a local guide in each city.

Everybody in our group was delighted with our local Baku guide and congratulated Chain Robbins, the group leader, as if the man had anything to do with it. Chain graciously accepted the unearned compliments.

Farah, which in the local language means "happiness," was petite with long black hair and dark eyes like many Azerbaijanis. She had a perpetual smile on her face. What a contrast from Moscow where we first met Svetlana and Skaya. All the men in our group were falling over each other to introduce themselves to Farah. Toxic Joe wanted to "date" Farah. I reminded him, "Man, you have a beautiful girlfriend back in Rahway, New Jersey (Joe had shown me her photos). If you touch Farah, you can forget about your girlfriend in the Garden State. You will rot in a Soviet jail for years. I am told that time goes slow in Russian prisons because the laws of gravity are different here."

"Okay, my friend, I am not going to do any such stupid thing. The truth be told, I love Lori and we will marry next spring," Toxic Joe told me.

Thus assured, we moved on to the next important issue. What to do with Deanna's package that we were unable to deliver to her friend Alana in Moscow? We wondered who would be the lucky Azerbaijani beauty to inherit the Victoria's Secret undergarments?

Then, as if prompted by a divine tweet (before Twitter), we both exclaimed simultaneously: "Farah!"

Our excitement was short lived. Farah was destined to own the American gifts but how could we approach her? KGB operatives lurked everywhere in numbers greater than those in Moscow. The Russians never fully trusted these far-flung republics and kept a close watch on them. Finding Farah alone would be very difficult, and talking alone with her even more difficult. How could we be sure that Farah herself was not KGB? "Looks can be deceiving," I told Joe.

Toxic Joe volunteered to approach Farah, but I advised against it. "Joe, you may be more clever with ladies than I am, but this situation calls for a more balanced and stable person with polished diplomatic skills." I sort of looked around in disbelief that I had just said this about myself.

"So I suppose that person is you?" he said in a slightly sarcastic tone.

I explained: "I think I can bond with Farah. The reason I think so is that I understand some of the words of the local dialect. For example, when we were riding from the airport to the hotel, we passed a bookstore. The sign in front read *Kitab Khana*, which in the Urdu language means a bookstore. Urdu, the language I studied in India when I was a small child, is very similar to Persian, a language many Azerbaijanis understand. You may recall that when I saw the sign, I instinctively read it aloud, and Farah looked at me in surprise and appreciation. I've learned through my travels that language can be a powerful glue for bonding."

"Dream on!" Joe exclaimed. "I am not even sure she heard you, but I do acknowledge that maybe you can use your meager Persian vocabulary to bond with her."

The stage was set. I would use my language skills as a ruse to approach Farah. This might give me the opportunity to speak with her alone.

Bus ride to the Fire Temple

The opportunity came sooner than we expected. Svetlana announced that evening at dinner that the next morning we would visit a very special place, the ancient Fire Temple, whose origins the experts were still trying to untangle. The visit would be the sightseeing highlight of our Azerbaijan trip. She said we were lucky. We would have three tour guides on this outing instead of the usual two. Because of the archeological significance of the mysterious site, we would be accompanied by a local tour guide intimately knowledgeable with the site.

Two guides always sat together. This time, however, there was a good possibility that the third guide, hopefully Farah, would sit separately. And maybe, just maybe, I could maneuver my way to sit next to her. This could give me the opportunity to bring up the subject during the two-hour bus journey.

Farah did take the empty seat, but alas, before I could act, another guy from our group beat me to the spot next to her.

My brain went into a more intense scheming mode. I would have one more opportunity to speak with Farah when we returned from the Fire Temple. I told myself, *I must maneuver myself into the seat right next to Farah.* Hopefully we would be away from Svetlana. Such a maneuver would not be an easy feat, as all the INTOURIST guides typically tried to sit close to each other.

After our tour (of which I remember nothing), while boarding the bus back to the hotel, I stayed close to Farah. I was not going to let anyone wrestle me out of the spot next to her. I stayed glued to Farah until I claimed the seat next to her. She had noticed this and seemed a bit irritated by my brutish behavior, but she still managed a smile and a subdued, "Hello."

This was my opportunity! Svetlana and Skaya were two rows in front of us. The rumbling noise of the speeding bus and my close proximity to Farah provided an ideal opportunity to talk about unspeakable things. I did not have any time to waste.

As soon as we started moving, I turned to Farah and said, "I am going to tell you something and hope you will keep it a secret."

Farah could not believe her ears! She said in a hushed tone, "You better not because we never have secrets between the INTOURIST staff and certainly not between the foreigners and the INTOURIST staff. Our work is strictly business. Please be careful. I don't want to get into any trouble. I know you can read the Persian signs here. Don't try anything funny!" As if as an afterthought, she then asked: "Are you crazy?"

"No I am not crazy enough to jeopardize your safety or mine. I just wanted to tell you that I brought some presents from America to give to someone. These are just normal gifts and souvenirs from America that I know are hard to find in the Soviet Union. All my gifts are for a lady because I am not very good at giving fancy gifts to men. Frankly, I am not sure if I will meet anyone more deserving than you to receive the package. You should know that I am carrying the gifts right now with me in my travel bag."

Her jaw dropped! "You are carrying expensive presents from America that you want to give to me? Why? I don't even know you." She continued, "Are you," she hesitated and said, "what is that word you Americans are so fond of saying? Is this BS?"

"No, this is not BS," I assured her. "It is a long story that we do not have time to discuss. I think Svetlana is staring at us, so why don't you look outside the window and point at some interesting feature, pretending as if you are telling me something about the landscape."

Farah smiled and said, "You are full of . . ." She never completed the sentence but chuckled at her own language prowess. She opened the window and pointed at an old unimportant building.

Svetlana was beginning to dose off in the warm Caspian air, providing me the perfect opportunity to discretely open the top of my shoulder bag to reveal the merchandise I was carrying. Farah took a quick peek at the L'Oreal cosmetics, the folded Victoria's secret undergarments, and the gleaming gold bracelet.

Her eyes widened, "Close it. Close your bag before someone sees and starts wondering what you are up to!"

"These are not illegal goods or merchandise I brought into the Soviet Union to sell for profit. I brought these to give to a friend of a friend in Moscow, but I could not find her. I don't know what happened to her. I don't want to take them home and be frisked by a US customs agent and be compelled to answer lot of questions. Would you please help me by taking them off my hands?" I pleaded.

"Okay, okay. Don't tell me anything else. It is still hot merchandise. You could get into trouble. And don't tell your story to anyone else. I will help you, but don't display these items. Many people in Russia would kill for these. Items like these are not available to us no matter how many rubles we have in our pockets."

"Thank you. I really want you to have these things for the simple reason that you are so beautiful!"

"Flattery will get you nowhere," Farah said with a big smile.

Svetlana was now sound sleep, but the burly bus driver was beginning to take an interest in Farah and me. He frequently glanced into the rearview mirror, which he adjusted many times. Nevertheless, we continued our discussion on how to transfer the precious cargo. I suggested to Farah that when we got back to the hotel, we should sit next to each other in the lobby. I would repack my backpack and place the items into a bag. When no one was looking, I would slide the bag toward her.

"Clever, but too risky," Farah said. "The lobby is crawling with KGB agents. Probably every fifth person in the lobby is KGB."

"So why don't we do this when there are only four people in the lobby?"

"Very funny, but the fourth one surely will be the KGB agent," she smiled.

"Okay, I have exhausted all my ideas" I said. "Do you have a suggestion?" I knew that having seen the goods, Farah wanted them and would find a way.

"Yes, I do," she said with confidence. "Are you staying on the second floor of the hotel? I believe this is where your group is supposed to be staying. If you are, I can come to your room, if only for ten seconds, and you can give the things to me."

"How is that possible? If I can't even slip in the merchandise to you sitting next to you in the lobby under the watchful eyes of the KGB, how would they let you go unwatched to the second floor and come to my room even for ten seconds?" I was aghast at her daring plan, one more reckless than even Toxic Joe could conceive.

"It is possible," Farah replied calmly. "The hotel is our base. We don't live there, but we have a changing room on the second floor. At the end of the day, the tour guides go up to the changing room where we have our own closets. We shed our INTOURIST uniforms, wash up, and change into our street clothes to go home. The whole process takes fifteen minutes. When we are back in the hotel later today, I will go to the changing room and I will finish a few minutes earlier than the other girls. I will then stop at your room and you can give me the presents. Make sure that you keep your door partially opened when I come in and don't peek outside. I will tap the door gently, come in, and you can give me the items. Afterward, don't come out and don't follow me."

"And you said I was crazy!" I was aghast at her plan. I could not close my mouth because my jaw had slid down to a record low. I complemented Farah on the thoroughness of her plan and commented, "Is this part of your tour guide training? Are you sure you are not KGB?"

"No, I am not. If I was, the KGB would be waiting for you when we arrive back at the hotel."

Toxic Joe was sitting right behind us on the bus and was watching every movement. Once in a while he would bump my back as if he had figured out something. His actions were very annoying.

We arrived back at the hotel precisely at five p.m. I rushed up to my room to get ready for the secret door knock ten minutes later. Joe stayed in the lobby. I had instructed him not to go up to his room and to resist the temptation of loitering around to get in on the action. Such interference could disrupt the whole process. As I was ascending the stairs I saw Farah from the corner of my eye starting to come up. We did not look at each other.

Precisely ten minutes later, I heard a gentle knock on the slightly ajar door. Without peering out, I opened the door another few inches. Farah slid in. We did not exchange a word. She opened her INTOURIST satchel and I dropped the goodies into it. She proceeded to exit, but before leaving, she turned around and planted a firm and gooey kiss on my right cheek and then the left. Very European. I felt the viscosity of the pigment transfer and the sweet sensation of her fresh lipstick. I looked in the mirror. There was a bright pink oval imprint at least three inches in circumference on each cheek. A forensic expert could have easily traced the pattern back to its rightful owner. I almost wiped them off before going down to the lobby where Toxic Joe was anxiously waiting to see if our mission impossible was accomplished.

I decided not to wash off the precious imprints until Joe could see them. This would be proof positive, the official stamp testifying that the goods had been delivered. I lowered my face and proceeded downstairs while trying to partially cover my face without being too conspicuous. At least two people saw the marks and made a gesture telling me that I had something on my face that needed to be removed. One lady laughed. Toxic Joe was watching. He went berserk. He started laughing like a hyena and pointed out my temporary tattoos to three others in our group who were still hanging around in the lobby.

"Mission accomplished," I declared and immediately proceeded to wipe off the pink imprints with tissues I had been clutching in my hands for the last few minutes. Joe was explaining the "lip frescos" to our three colleagues and they were going hysterical. I also noticed that several other people were laughing, too, no doubt including every fifth person who must have been the KGB agents that Farah alerted me to. They had no idea what had transpired but they were sure of one thing: the crazy Americans were up to something again.

The next morning we were scheduled to take a bus to Tbilisi, Georgia, the next destination on our tour. Svetlana would ride with us, but unfortunately we were leaving Farah behind. Farah came to the bus terminal to say good bye. The group entered the building single file and I dragged my feet so I could be the last in line. At the doorway I turned around and waved to Farah who was standing at a distance greater than necessary. I looked at her as she blew me a kiss. The kiss did not deposit any pigment on my cheek, but it left a lasting imprint in my mind!

CHAPTER 24

STOLICHNAYA EXPRESS
TO GEORGIA

Behind the Iron Curtain – III

The bus trip from Azerbaijan to Tbilisi was a long four hundred kilometers. We were advised before the trip began that we were in for a long ride. This did not matter; the visitors from America were excited. For the first time we would be seeing the Russian countryside. Until now, all our views had been from small windows at +/- 30,000 feet.

On this trip the state tourist agency chose a local Georgian guide from Tbilisi, the capital of Soviet Georgia, to accompany us from Baku, Azerbaijan to Tbilisi.

The Georgian guide's name was Eliso, which we figured to be the Georgianized version of Elizabeth.

Eliso was slightly on the heavier side—not fat, but just a bit overweight—possibly a result of having been raised on the wonderful rich dairy diet so abundant in her native land. She was also perky and agile. Eliso was about thirty years of age. She seemed to have a warm disposition, so I guessed that she would be an easier target for "softening up" compared to the other hard-edged INTOURIST guides that accompanied us on our trip around the Soviet Union. Eliso displayed an immediate interest in mingling, but Svetlana's stern gaze told her to steer clear of the rowdy foreigners.

The potential discomfort of the seven-hour-long trip was not lost on Toxic Joe. We would not have another road trip on our tour that would take us through the rural countryside. Joe wanted to make sure every minute of this journey was gainfully spent and enjoyed. Before starting, he made sure to load up on needed supplies and entertainment accessories for the trip.

The burly Georgian bus driver, who was introduced to us as Djaba (pronounced "Jaba"), seemed friendly. Whether or not he understood what was being said, Djaba instinctively nodded in agreement—probably a good habit to learn in Stalinist Russia. Certainly he had found a good job. My guess is that hauling around all kinds of strange people from all over the world who spoke different tongues sure beat working on a collective farm or in a state-owned factory.

By seven a.m., the bus to Tbilisi was on its way. Svetlana introduced Eliso to our group and assured us that we would be in good hands. Svetlana said that Eliso was very knowledgeable and knew the city of Tbilisi "like the back of her foot," which we interpreted to mean "the back of her hand." In her nervous attempt to use American slang, Svetlana sometimes confused parts of the human anatomy. A few people snickered but most kept quiet.

Svetlana patiently introduced each one of us to Eliso. When doing so she occasionally peeked at her 'cheat sheet' list of names and nicknames.

When it came to Joe, she said to her new colleague, "Eliso, this is Toxic Joe." A roar went through the bus. Realizing she had said something not quite right, she muttered, "I know there is something wrong about what I just said, but I don't know what. I will find out later."

"Hi Toxic Joe, nice to meet you," Eliso said politely in halting English. Another roar went through the bus, this time exceeding the ninety-decibel Russian noise exposure limit. Even Svetlana contributed a little to the time-weighted exposure noise dose.

As he drove, Djaba started humming what was probably a Georgian folk song. Svetlana updated her travel logs. Eliso peeked inside the ice box and the picnic baskets to verify that there were enough snacks, including bottles of sweet Russian soda, to last through the entire journey.

All of a sudden, a commotion broke out in the back of the bus where Joe was sitting with two other troublemakers. Shouts of "Wow!" and "Awesome!" and "What a Guy!" filled the bus. Djaba was so startled that he almost lost control of the vehicle.

Toxic Joe had just unveiled the cargo he was carrying in his bulging backpack in the rather dramatic fashion characteristic of him. Inside his bulging luggage were a dozen shiny bottles of Stolichnaya (Столичная) Vodka. The gathered delegates started shouting "Stoli! Stoli!" as if cheering for a countryman near the finish line of an Olympic one-hundred-meter dash.

Djaba looked into the rearview mirror in happy disbelief. If he had three hands, he probably would have pinched himself to see if he was dreaming. *Am I having a dream? Did I doze off? Could I be in the American Georgia?* he wondered. Later he told us, through Eliso, that he had heard that there was a Georgia in America. He always wanted to go there. He had heard that in American Georgia you were allowed to bring vodka on a bus. Djaba was probably wondering how this could be happening in the Union of Soviet Socialist Republics. After all, foreigners were drinking alcohol on his bus while government officials, Svetlana and Eliso, were present.

He peeked in the rearview mirror again. The ethanol-based nectar was being passed from one seat to another, and he was on no one's list to savor this elixir of the gods. Djaba's increased frequency of peeking through the rearview mirror, an unsafe movement, made me nervous.

Svetlana searched for her megaphone and shouted, "Gentlemen, please put the alcohol back in the bag. You are not allowed to carry alcohol inside the bus, let alone drink it. This is illegal. We will all be in big trouble if you continue. Please be wise and stop!"

No one seemed to be in a mood to be wise except our tour leader Chain Robbins, who tried to echo Svetlana's sentiments. Chain's sage words were lost in the clinking, cheering, and laughing as Toxic Joe poured crystal clear fluid into the tiny glasses he had carefully packed into his travel bag. Svetlana tried repeatedly to warn us, but to no avail.

Professor Dietrich felt the urge to mediate. (At the time of our technical exchange, Dr. Dietrich A. Weyel, CIH, was an assistant professor of occupational hygiene at the University of Pittsburgh.) "It is okay, Svetlana, no harm will come," Dietrich advised. "No KGB is on this bus. We know this because everyone in the bus is a known entity. When we arrive in Tbilisi, we will all behave and no one will ever know."

He continued, "Have you ever heard of a famous city in America called Las Vegas?"

Without waiting for an answer, he explained, "The whole town is about gambling, drinking, and entertainment. Anything and everything goes, and everything you do in Las Vegas is instantly forgiven by everybody, even by the government."

At this, Svetlana yielded a most disapproving look.

"By the way, do you know what they say about Las Vegas?" Without waiting for an answer from Svetlana, Dietrich answered his own question, "Whatever happens in Las Vegas, stays in Las Vegas."

Svetlana shook her head in disagreement. "What happens inside a Soviet bus will not stay inside a Soviet bus," Svetlana assured Dietrich.

Finally Svetlana threw up her hands in frustration, released an audible groan, and slumped in her seat. It was pointless. She had no options left but to ignore it all and tolerate the revolt for the remainder of our road trip to Tbilisi.

The bus driver Djaba seemed agitated and fidgety but for a different reason. As he kept looking through the rearview mirror, this nectar of angels was being splashed around and he was getting none of it. The sweet and pungent aroma of pure ethanol was killing him. We felt his pain, but we were not going to give him alcohol while he was driving. If a report got out, the poor fellow would surely lose his livelihood. I didn't think he would enjoy working on a collective farm. More importantly, no matter how rowdy and reckless our group was at times, we were, above all, safety professionals. We would not allow a worker on the job, a hazardous one at that, to be compromised by drinking alcohol, exceeding the speed limit, and endangering his precious cargo and himself in any way.

Soon Toxic Joe ended Djaba's agony. Joe walked up to the front of the bus and discretely slipped a large bottle of Stoli into Djaba's travel bag, which was sitting by his side. His agony ended, Djaba produced a smile as broad as the Georgian steppe.

"Thank you, friend," Djaba acknowledged the gift in clear English. This statement utilized fifty percent of his hard-learned English vocabulary in the process.

"My pleasure, buddy," Toxic Joe said, tapping Djaba's shoulder and returning to his seat in the back of the bus. The driver nodded his head and picked up speed in excitement. He modulated back to the correct speed limit when he realized his mistake.

Modern palm reader

Things during the ride seemed to be settling down even though the Stoli was still flowing. People in the bus were behaving. Suddenly, Joe turned to Eliso, who was sitting near to him, and said: "Eliso, do you know that my friend Jas is a palm reader? His heritage is Indian and he shares the interest of his countrymen in telling fortunes. Because he is a PhD scientist, he is more accurate than most fortune tellers," he continued. "Jas is remarkably accurate with women." (This happens to be true.) "Should I ask him to come over and read your and Svetlana's palms?"

Before Eliso could say anything, Svetlana ranted, "No, it is not permitted! Better that Jas stay where he is. Absolutely no reading of any body parts in this bus. Fortune-telling is illegal in the Soviet Union. We Soviets do not believe in such things. Those are decadent and exploitive bourgeois pastimes. Such activities are not permissible in Russia and certainly not inside a government vehicle or on government property. There is nothing scientific or technical about staring at someone's palm." Svetlana's agitation was genuine.

At this point I could not stay quiet any longer because my judgment and skills were being challenged and ridiculed. I needed to defend my trade.

"Okay, Svetlana, I am not begging you to let me read your palm, but let me tell you that modern palmistry is a science. It involves personality judgments and psychology. Back in the USA, people line up when they realize I can read their palms. The fact is that most of the time I surprise people with the accuracy of my judgment of their personalities and behavior. I can even tell their future with a remarkable degree of accuracy. Let me tell you—at our last professional conference, I studied the hands of dozens of highly educated people, including scientists and engineers. They let me probe their palms like obedient little kindergarten kids. All of them were surprised with the accuracy of my descriptions of their past activities. And what I predicted about their futures seemed plausible. Let me tell you about one pretty young lady. She was of Eastern European heritage. Her name was Chito, which she told me meant "bird" in her native tongue. I looked at Chito's left palm because the left hand is the one you read for ladies in modern palmistry."

At this Svetlana made a funny face, but I continued.

"I told Chito all about her hobbies, her likes and dislikes, what kind of men she had been involved with in the past, and what kind of men were in her future. I told Chito that against her better judgment, soon she would fall in love with an older man, a man with a little bit of an accent, who was probably a professional in the health and safety field. Additionally, he would be smart and funny."

Several chuckles sounded around the bus.

"Chito finally realized what I was saying, and in one jerk, withdrew her hand. She looked at me, smiled, and said, 'You are such a rascal.' Then she said to me that I was right on target on everything except for that last piece of bullsh!t."

This brought a smile to Svetlana's face, which she tried to snuff out in an unsuccessful, though brave, effort. Eliso could not exercise any such restraint and laughed.

I kept hammering away in a vain effort to read Svetlana's palm. I said to her. "Svetlana, this is all in fun. I don't exploit people. In fact, when I go to our professional conferences, I am generally surrounded by some of the prettiest ladies waiting in line for me to read their hands."

"You, yourself proved my point," she said. "Your palmistry 'science' has no redeeming value. It is all BS, using your own favorite phrase. But congratulations! Enjoy your palm reading ventures at your next conference in America. Here in the USSR, don't expect Russian ladies, pretty or not, to line up and extend hands to be read and listen to this bull.." She stopped, realizing she was not supposed to indulge in such informal discourse. Such language use was prohibited under INTOURIST rules.

I was elated because I just saw a clear ray of hope. Oh my God, the hard core comrade was actually enjoying the decadent capitalistic bull—

Soon everyone on the bus wanted me to read their palm, which allowed me to draw upon huge reserves of BS that I had been carefully storing away.

I couldn't believe it. Even Eliso briefly extended her hand. Svetlana ignored this and looked away. What in the world was happening here? The Kremlin's walls were crumbling. The capitalistic pathogenic organism was spreading like the Asian flu.

I obliged Eliso and told her she had much foreign travel in her future. She asked if the destination looked anything like America. I nodded in agreement. This pleased her immensely.

By now the lights of Tbilisi were getting closer. Finally we had reached our destination. Svetlana and Eliso checked us in at the Tbilisi hotel and stayed with our group until dinner was over.

Before leaving, Svetlana turned toward me in the dimly lit lobby and murmured,

"Whatever happens inside a Soviet bus, stays inside a Soviet bus."

"Absolutely." Our lips were sealed, I assured her.

"You guys are nuts." She couldn't help saying another slang word she had learned on this trip while walking away.

"Yes, and we are proud of it," I replied.

Svetlana kept walking. I could hear her giggling from fifty feet away.

CHAPTER 25

EXCHANGING GIFTS CAN
BE HAZARDOUS

Behind the Iron Curtain – IV

Kiev, Ukraine, was our next stop.

Located along the Dnieper River, Kiev, (Київ in Ukrainian) is the capital and the largest city of Ukraine. The city is one of Eastern Europe's most important industrial, scientific, and cultural centers.

The city's name is believed to derive from Kyi, one of its legendary founders. Kiev is believed to have existed as a commercial center as early as the fifth century. It was a Slavic settlement on the trade route between Scandinavia and Constantinople.

Kiev prospered again during the Russian Empire's industrial revolution in the late nineteenth century. From 1921 onward, Kiev was the most influential city of the Ukrainian Soviet Socialist Republic, and from 1934, its capital.

During World War II, the city suffered significant damage, which they honor with many war monuments. Kiev quickly recovered in the postwar years, remaining the third largest city of the Soviet Union. Following the collapse of the Soviet Union and the Ukrainian independence of 1991, Kiev has remained the capital of Ukraine.

We arrived in Kiev late in the evening. The KGB was never very far away. Sensing their presence was not just our imagination. We could feel it. We could smell it. We could see it.

Not used to being watched, the American delegates were not always careful about what they said in public—to each other or in the privacy of their hotel rooms or within earshot of the INTOURIST guides and the bus drivers. The drivers did not know English, but that did not mean they were not trying to understand what was being said. Who was to say whether the driver himself was not a KGB agent masquerading as friendly Djaba?

While out for a walk on our first night in Kiev, four of us troublemakers soon bumped into a group of young Russians, four women and three men. All but one were college students, bright and articulate with an amazingly detailed knowledge of the United States. Their knowledge included facts related to geography, politics, literature, and yes, American music.

The young Russians wanted to know which part of America we were from, what brought us behind the Iron Curtain, and where we were staying in Kiev. They also wanted me to assure them that I was not Cuban. I was never sure of the reason for their intense concerns about my suspected Cuban heritage. (I am not Cuban and have never even visited Cuba.) Perhaps it had something to do with the confidentiality of the transaction we were about to conduct. Whatever it was, my friend Professor Dietrich assured them that "the Cuban fellow posed no security risk." Most of all, the young Russians wanted to know if we were interested in a souvenir swap. We told them we were.

Igor, the oldest student in their group, suggested that we discuss such matters in his apartment, which was only a few blocks from where we were staying. It would be better, however, if we did not walk with them but instead followed them at a respectable distance. We complied as Igor led us to an old, dimly lit apartment building. I thought Igor had a job, but he did not tell us where he worked or what he did. Igor shared the little apartment with some women. The prettiest one, named Natasha, acted as the hostess. Igor and Natasha were very hospitable. They served us local champagne and chocolates (chocolate consumed with champagne was a popular local treat).

Discussion soon turned to a souvenir exchange. What treasurers did we bring from America, and what could Igor and his party offer us in exchange? We told the Russians what we were looking for. We wanted to know what they had that we could take home without getting arrested at the Soviet customs in Moscow.

The Russians were ready to deal. While most of us were focused on the famous painted Russian dolls and heavily lacquered wooden boxes, our Russian trading partners had a more carefully thought-out priority list, which included Levi's Jeans, cosmetics, ladies' undergarments, and any past issues of *Time* magazine.

They were hungry for any news from outside the Soviet Union. We had a generous supply of Levi's jeans because before our trip, everybody in the US had told us that Levi's were what every Russian wanted and to forget about anything else. To a large extent, that was true.

As for the cosmetics, we could do reasonably well with those items. At least a third of our delegation was comprised of professional American women who we suspected would reluctantly part with some of their cosmetics if we begged and pleaded enough. I didn't think that it would cost us too dearly.

As for old news magazines, unfortunately it was impossible to comply with our host's wishes. None of us could have predicted that the heaps of old magazines some of us had back home gathering dust would be treasured documents in the land of Karl Marx, Joseph Stalin, and Vladimir Lenin.

One would think that none of the items they were seeking would be particularly sensitive or clandestine except possibly the *Time* magazines. The KGB may have thought differently.

From our side, Toxic Joe was not satisfied. "We are not getting a fair deal," he whispered in my ear.

"Why is that?" I inquired.

"They are getting blue jeans and expensive cosmetics, and all we are getting are those goofy looking wooden dolls in exchange," he complained.

"Okay, what else do you want from the Russians?" I asked Joe, who was still whispering in my ear and drawing curious looks from others in the room.

Toxic Joe further lowered his voice so I could barely hear him. "One of the Russians told me earlier that he has a used Russian military uniform that one of his relatives gave him."

Apparently the relative had served in Afghanistan for three years when the Russians were trying (unsuccessfully) to put down an uprising in Afghanistan, which curiously enough found the Americans supporting local freedom fighters. This was long before we Americans had ever heard of the Taliban and Osama Bin Laden.

"Are you crazy?" I asked, loud enough to alarm everyone in the room. I resumed in a whisper. "Listen, Joe, this Russian army uniform exchange will probably cost you more than three pairs of Levi's, but more importantly, the Russians will never let you take a uniform out of the country. They will confiscate it and quite possibly, they will confiscate you along with the uniform."

"Don't worry," he assured me. "I can round up plenty of Levi's. Everyone in our group is carrying a sh!tload of the jeans, and as far as getting the uniform out of Russia, do you see that mousy little Russian guy sitting next to Natasha?" Without pointing him out, Joe drew my attention to the nerdy fellow with the rimmed glasses in the corner. He looked exactly like the famous American actor, writer, and director Woody Allen. Joe whispered to me, "Woody told me that it would not be a problem. Just tell the security man at the airport that we are a US government unit and the uniform is a memento of our US-Russian friendship and peace message that we are carrying back to America."

This line did not sound very convincing to me, but I said to Joe, "Okay, we will discuss this later." I wanted to end our side conversation and rejoin the others. We compiled our list and promised to meet again the next night after nine p.m. in the same apartment to execute the exchange.

The next evening we followed our plan. The 'jeans-for-dolls' exchange was made. With that completed, Woody Allen then handed over to Toxic Joe the Russian uniform wrapped tightly in several layers of rough brown paper without saying a word.

By the end of our meeting, a significant US-Russian trade deal had taken place inside Igor's tiny apartment even though not a single coin of any currency was involved. The method used was the old-fashioned barter system from simpler times.

Once business was done, Igor immediately insisted that we had to celebrate with another round of champagne and chocolates. We wrote friendly notes of thanks to each other and then said *dosvedanya* before we returned to our hotel.

Fortunately no one followed us. The risky secret transaction had been executed with precision. No one was lost or injured.

Air vents have ears

We were so pleased with our deeds that, as usual, we decided to gather in Professor Dietrich's room for a toast to our courage and success using the finest vodka that money could buy. We invited as many of our fellow delegates as we could find to join us. Some of them had already gone to bed, but many others came to celebrate our trade mission.

Toxic Joe described each transaction with the flair of a Sotheby's antique auctioneer. The merchandise that Joe first talked about was the exchange of a pair of Levi's for a lacquered wooden box coated with red pigment that probably contained toxic lead oxide.

With great relish, Joe then described the exchange ceremony that featured two pairs of Joe's Levi's for an authentic Soviet military uniform worn by a Russian war hero during the disastrous Afghan war—Russia's Waterloo. Joe also described how the Woody Allen look-alike told him to smuggle the uniform through customs without getting arrested.

Someone then looked up at the air-conditioning vent and said, "Hey everyone, look up; someone may be listening to all of this. Don't you realize? Air diffusers are ideal places for planting listening devices. Maybe the damage is already done."

"Never mind," Professor Dietrich declared. "What are you worried about? We have no information that would be of interest to the KGB. All we have is a lot of BS. Don't forget, guys, we are Americans. Like most of our countrymen, we have no knowledge of other countries, Russia included. We don't even know the names of most of the cities on our tour. Hell, most of us don't even know the capital city of Canada."

Then, raising his left arm as if he was teaching his class back at the University of Pittsburgh, the professor asked to see a show of hands of those who knew which city was the capital of Canada. While several in the group were still scratching their heads, two of the geniuses yelled "Toronto."

"No, it is not Toronto, you morons!" He hesitated, then repeated his words while tilting slightly toward the vent and said, "You guys! I, Professor Dietrich, have just insulted two of our most experienced CIH's who incorrectly guessed Toronto, not Ottawa, as the capital of Canada."

Professor Dietrich then looked straight up at the vent, cupped his hands, and shouted, "Hey, up there, whoever you are and wherever you are, if you are listening, I would appreciate you sending me a cup of hot coffee at five o'clock tomorrow morning. We have an ungodly early wake-up call tomorrow. I cannot function that early without some caffeine. A cup of coffee would be greatly appreciated. *Spasiba* (thank you)."

All of us looked toward the air vent and almost in unison shouted, "*Dosvedanya* (good-bye)." It was time to go to bed for the early wake-up call.

Service with a smile

The next morning at breakfast, Professor Dietrich told us an incredible story. He said that at five minutes after five a.m., he had heard a knock on his door. He opened the door, and there was a man in a fancy red uniform holding a shiny tray with a steaming cup of black coffee and a biscuit.

"With the compliments of the hotel, sir." The man in the red uniform placed the tray on Dietrich's desk, bowed, and left.

We were stunned. A debate ensued. Was the professor just piling it on higher and deeper again? Or did this really happen?

Dietrich stuck to his story. "You can believe it or not, but you cannot deny the fact that only I had the five a.m. delivery of coffee and biscuits. Did anyone else get the wake-up coffee? Please, a show of hands." The professor was being professorial again. No hands went up. Dietrich offered to take us all to his room to show us the empty coffee cup and the uneaten biscuit on the tray as proof.

"Still don't believe me? Let us ask the guy in the red coat who else he served."

We were finally convinced that the professor was telling the truth, but how did this happen? Wasn't it nice that he got served a beverage instead of an early-morning interrogation by the KGB? Could it be that KGB has a heart and sense of humor after all?

The time was now eight a.m. Just outside, in the driveway, the burly Russian bus driver was getting impatient. Svetlana waved her hand, "Ladies and Gentlemen, it is time to go."

CHAPTER 26

BIG FAT ARMENIAN WEDDING

Behind the Iron Curtain – V

Yerevan, Armenia

At each of our tour stops, our routine included either an obligatory technical exchange visit to a governmental agency or a sightseeing tour.

Sightseeing was usually the highlight of our activities because the technical exchange activities were not very informative. Russia did not have much applied industrial hygiene at the time of our visit, which meant there was usually not much opportunity to exchange information related to our professional interests.

Every now and then we found areas of common interest, but on those rare occasions we would hear a research presentation by one of the Russian academicians. More often, our group was briefed on medical or dental clinic practices when we were actually searching for details of industrial hygiene practices in Russian industry. As a result, large parts of our technical discussions were usually consumed by lengthy greetings and elaborate, but pleasant, introductions.

As previously reported, our movements were tightly restricted, especially when we had no assigned guides. Consequently, we had little to do during the evenings other than drink pepper vodka in Professor Dietrich Weyel's hotel room.

Friday in Yerevan must have been our lucky day. We learned through keen observation that there was a big wedding at the hotel where we were staying. Given that Yerevan was the capital city of Armenia and we were staying at one of the nicest hotels, this was going to be a big deal. What's more, this adventure was going to be right in our own backyard. There was no need to arrange illicit transportation, no confusion over directions, and most importantly, no dodging the KGB, unless of course the KGB had infiltrated the wedding party's family.

The prospect of experiencing an authentic Armenian wedding was very enticing. The only problem was that none of us had an official invitation. Professor Dietrich, being very knowledgeable about global customs and protocols, recommended that a couple of us should try to crash the wedding party, but we should wait until the invited guests had consumed a healthy dose of ethanol.

I had heard that crashing a Russian wedding for Westerners was easy. Russians are friendly hosts, especially when under the influence of Stolichnaya Vodka, which we knew would be flowing like the Volga River in summer. Moreover, I felt confident that with a little stretch of the imagination (lubricated with ethyl alcohol), I could pass as an Armenian émigré who had settled in Southern California like many other Armenians.

The other two party crashers, Professor Dietrich and the Texas oil man Coleman Hahn, looked less Armenian, but they could have been Ukrainians who had settled in Armenia. This meant that no strenuous exercising of our imaginations, beyond our usual mastery of BS, would be necessary.

Crashing a party – Yerevanian style

Professor Dietrich was a fascinating character. He was of German descent, stood over six feet tall, and wore a neatly trimmed beard. His casually fashionable clothing (called "business casual" in some parts of the world) and his equally comfortable manner gave him easy access to people. Dietrich was the kind of guest every bride desired to have at her party. He was very engaging.

Hahn, who worked for an international oil company, was shorter and less physically imposing than Dietrich, but you could never ignore him. He would not allow it! He was a consummate talker, engaging and interesting in his own right. He would never hesitate to walk up to an attractive woman and introduce himself. The man was fearless when it came to interacting with women.

"Hi, I am Hahn, and what did you say your name was?" he would say even though at that point no one had yet told him their name. It was immaterial. The name formality over even before it was begun, Hahn would immediately ask for a date. I was raised in a very different environment, so this behavior intrigued me. And I admired Hahn in a weird way. Maybe I was envious of his success rate.

What a crude approach, I would think. *How could he possibly expect any positive responses from this rough technique?* Strangely, though, it worked. The proof was easy to see. The man was never without an attractive date. You couldn't help but acknowledge his extraordinary talent.

"Your approach seems outright crude. Do you expect any self-respecting woman would leave a party with you when you barely know her name?" I challenged him.

"Oh, you poor sheltered man from the country. First of all, why do you even need to know someone's name, especially here in Russia? I mean, what's in a name? Haven't you read Shakespeare?"

Then he continued, "Obviously, you haven't mastered statistics, despite your PhD in chemistry. It is all statistics. If you approach fifty women in the short time you have at a party, odds are in your favor that eventually you will make a good connection."

It was not my way of socializing, but it made sense for someone like Hahn.

We found the wedding reception room and strategically waded into the jubilant crowd when the musicians started playing a popular Armenian wedding song. It was called:

"Ha janum, elek jrag varetsek"
("Dear one, go and light the lantern")

It was beautiful music. I wanted to dance, but I didn't have a partner. Hahn and I slowly moved closer to the dance floor. Many stares followed, most just out of curiosity, but some showed a mixture of amazement and even admiration. After all, we were well-dressed and looked very respectable.

A middle-aged man dressed stylishly with somewhat of a "pimpish" demeanor walked towards us, extended his hand for the 30 PSI handshake, and said: "Welcome, gentlemen. My name is Azad." (*Azad* means "free" in Armenian.) "I am a distant uncle of the bride," he said with some hesitation, telling us in 'guy' code words that he was a phony uncle, like many phony uncles all around the world.

Digression Alert: I have learned that the "uncle" connection is frequently abused to try to mask illicit relationships. I have come across many phony uncles. One time at the annual professional AIHA conference, one of the older guys, whom

I had known for many years, introduced me to an attractive young lady, who I estimated to be about half his age, as his niece Nina. He said to me, "Nina is interested in science, and especially chemistry. I brought her with me to the conference so she could meet some smart chemists like you and get some advice about her future career. There are several branches of chemistry. Do you think Nina would like physical chemistry—your specialty, I believe?"

I had instantly determined that Nina was not chemistry or physics material. She was there to play house with the bogus uncle, and the whole thing was a charade. I decided to play along and oblige the man whom I had known for a long time. I admired his technical prowess. I told Nina that I could talk with her at breakfast the next day before the technical sessions started. The next morning Nina and I had a nice conversation over breakfast. I quickly deduced that she wasn't brilliant in the scientific sense, but she was pleasant and cute. As I had already determined, Nina had no interest in chemistry, physics, or biology, although I could have been wrong about her interests in anatomy.

After Nina departed, I kept thinking, What a phony uncle and a dumb one, too. He had forgotten that several years earlier I had met his lovely wife and his real niece Nina, who was an intelligent young lady who seemed destined to be a medical doctor.

Back to the wedding and Uncle Azad.

Azad said, "Please feel at home. I know that you are not on the bride's guest list, but you are welcome anyway. You are our foreign guests." He then called over a waiter and had him bring us some drinks. Turning to Dietrich, he asked, "So, where have you gentlemen come from, and what brings you to this corner of the world?"

Azad was quickly impressed with Dietrich's explanation of our mission and our credentials.

Uncle Azad spoke good American English (he had lived in California for two years, he told us). He spoke in an authoritative voice like a military officer, which he was not.

The hall was full of attractive women, who outnumbered the men. This did not escape Hahn's attention. He wanted to demonstrate his mastery of statistical sampling immediately, but Professor Dietrich and I restrained him. We did not want to get kicked out of the party as soon as we had arrived. Additionally, being patriotic, we did not want to cause a diplomatic scandal by bringing shame and bad publicity to the United States of America.

We decided to stick close to Uncle Azad, who seemed to have his own following among the ladies. Sooner or later, we figured, he would introduce us to one of his beautiful friends.

"Why wander around?" Dietrich whispered. "Just stay close and be nice to Uncle Azad."

"Yes, let's stay with Uncle Azad," I agreed.

Azad excused himself for few minutes. A short time later he returned with two stunningly beautiful ladies who he introduced to us as "Armani" and "Dalita." Interestingly, Dalita meant "virgin" in the Armenian language. Hahn and I admired Dalita's name. *How clever,* I thought. *Even if you are not really a dalita, such a name is a great advertisement for purity.*

Armani, the taller of the two, was lighter in complexion and looked Russian or Ukrainian. She dressed as if she came directly from a fashion studio in Milan, Italy. She was poised and sophisticated. Unlike Dalita, who was more down-to-earth and approachable, Armani had an aura of detachment and mystery. At the same time, her eyes were sending clear signals—if you could gauge the wavelength, that is—that she was open to an adventure if one possessed the courage to cross swords with Uncle Azad.

Azad told us that the women were his "distant cousins." It seemed that just about everything in Uncle Azad's life was distant. Perhaps that is a successful approach if you have Azad's personality and style.

Armani caught the professor's fancy. He stuck to her like a distant uncle. I overheard him telling her how she reminded him of one of his brightest students back at the university. Hahn and I tried bravely to subdue our expressions as Professor Dietrich was piling on the flattery, but we failed. Finally I could not help yelling to the professor, "Dietrich, you are burying the beautiful lady under your PhD (piled high and deep)."

Both women giggled without understanding the pun. Dalita said, "That sounds like colorful American slang. What does it mean?"

"Maybe later," the professor said, obviously feeling a need to postpone the PhD discussion.

Professor Dietrich had now totally turned his attention to Armani and paid no attention to my conversation with phony Uncle Azad.

Hahn, on the other hand, made it his mission to correct Dalita's mispronunciations of some of the most celebrated four-letter American swear words as if it was his patriotic duty, which being from Texas, he always took seriously.

The band played another beautiful and haunting Armenian song. It sounded sad in contrast to the previously joyful song. I asked Dalita why they were playing such serious music at a wedding. Dalita explained, "The song they are playing is, 'Mayrik, du mi la' ("Don't you cry"). In Armenia, a wedding is also a sorrowful occasion. The bride will be leaving her home, her parents, and her childhood friends. From that time on she will live with someone who is essentially a stranger and her new home could be quite far from her parents. Such partings are very sentimental."

I could identify instantly with that sentiment. Growing up in India, I vividly remember such occasions when there was a woman's wedding in our own family. This was a traumatic event in the old days when distances were really far off and communications were limited to snail mail or couriers. I told Dalita that the song was very powerful and made me sad, too.

Time was quickly passing, and it was getting late. I wanted to go back to my room, as we had an early wake-up call. I wanted to grab at least a couple of hours of sleep because I could be called upon to assist our leader Chain Robbins with crowd control on the next day's journey. Because of my responsible reputation, Chain had appointed me as his unofficial deputy. This position carried no privileges other than sometimes I could sit next to one of the INTOURIST guides, a privilege usually denied to all other delegates except Chain Robbins.

I said a warm dosvedanya to all, including Uncle Azad. He gave me a big hug, which I thought was unnecessary, but I reciprocated with an equal but less warm hug. Hahn also wanted to call it a day, but he stuck around in the hope that maybe he could turn on his charms with Dalita once I was out of the picture.

The professor had no intention of going anywhere as long as Armani was there.

What did you do to my sweet girls?

The next morning we woke up bright and early to prepare for our trip back to Moscow. After packing our belongings, we went downstairs to get breakfast. As soon as we sat down, I saw this hulk of a man coming toward us in an extremely threatening manner. It was Uncle Azad. Paying no attention to Hahn or to me, he went straight to Professor Dietrich, stopping only a few centimeters from his face. Azad pointed a finger at Dietrich and thundered: "What did you do to my sweet girls?"

Dietrich played dumb, which was an interesting challenge for the professor. Trying to act innocent, he said, "What sweet girls?"

"Armani and Dalita," Azad said as he raised the decibels and repeated, *"What did you do to my sweet girls?"*

Dietrich stepped back a little and said, "I didn't do anything to your sweet girls. They were with you the whole evening."

"*No, they were not!*" Azad thundered.

Dietrich decided to stick with the story he had concocted. It was too late to waver. He did that in a convincing manner, having mastered the technique from being a professor who taught behavioral safety as one of his subjects.

Dietrich asked Uncle Azad to sit down and said, "Azad, I don't know where your girls went. Hahn and I left the party when you went into the bathroom and you did not come out for a very long time. We thought that maybe someone helped you to your room because it looked like you had too much to drink."

Dietrich continued, piling it on high and deep. "We waited, but when you did not come back, we thanked the bride and said goodnight to Armani and Dalita. And then we left. You can ask Rehana (the bride). The girls were not with us when we left the party."

Dietrich's BS seemed to be working. Seeds of doubt were now deeply planted into the phony uncle's brain. Azad was thinking, which may have been unfamiliar territory, maybe the Americans were right. He did have a lot to drink and he did go to the washroom and maybe someone helped him to his room. He did not remember.

Azad seemed resigned to it all, shook his head in frustration, and said, "Okay, fellows, maybe you are right, maybe not. Yes, I did have a lot to drink. I still don't trust you guys. You are fortunate that I like Americans, but don't push your luck! We still have gulags here."

"Thanks for the advice; we will remember," we all said in unison. "And thanks for the great time last night. Be assured we would never harm your nieces. They are so sweet and innocent."

Obviously, Uncle Azad did not believe our characterization of his nieces. We shook hands and quickly departed.

The bus to the airport was ready to roll. We hopped in and were off.

On the bus, Dietrich, while sitting behind me, whispered, "Uncle Azad's sweet nieces are really not that innocent. They were wild and out of control, and I don't think they give a damn about their distant uncle."

I already suspected that.

CHAPTER 27
ROMANCING THE STONE

It was our last day in Russia. The next morning our group was scheduled to fly out of Moscow to New York City, after which we would fan out in all directions across the USA. Our INTOURIST guides planned a farewell dinner for our delegation on our last night in the Soviet Union.

At our request, Svetlana, our main INTOURIST guide, also arranged a shopping trip to the famous GUM superstore in Moscow before our farewell dinner. Svetlana, who started the trip as a hard-nosed task master with an "it-is-not-permitted" attitude, was now a mellower figure, perhaps resulting from three weeks of haranguing, harassing, and brainwashing by her rowdy American visitors. I think we wore her out.

Facing Red Square, the GUM (ГУМ) store is an historic Russian landmark. Built in the nineteenth century, the super-giant shopping mall has been described as the:

State Department Store
Main Universal Store
Harrods of Moscow

The iconic store was rumored to fulfill the fantasies of even the most jaded shoppers.

Legitimate souvenir hunting was at the top of the list for many in our group. Having engaged in some illicit trades in a dimly lit Kiev apartment, we could finally openly buy exotic Russian goods with the help of the "greenback" (the US dollar). You couldn't purchase even a bottle of vodka with a pocketful of rubles in those days. Besides, foreigners just weren't allowed to own the local currency. You had to have what was called a "hard" currency, which meant US dollars, or a major Western European currency— German marks, English pounds, or perhaps Japanese yen.

Once inside the gigantic store, everyone headed in different directions. We quickly noticed that there was not much to look at. This should not have come as a big surprise to anyone who read newspapers, watched the news, or was even slightly aware of the world outside the USA, which unfortunately excluded a significant proportion of our fellow citizens. To most of our delegates, which included educated professionals from some of the top academic, research, and industrial institutions in the world, Russia's shortage of everything should not have been a shock. We had seen empty shelves, darkened hallways, and shuttered boutiques in stores elsewhere on our tour of Russia.

Nevertheless, the scene at the GUM store was outright eerie. The paucity of goods was sobering. You had to wonder, this was the crown jewel of Russian consumerism. Moscow was their largest and most important city. What happened here? World War II had been over for thirty-five years, allowing plenty of time for store shelves to be restocked with some decent merchandise. Alas, there was nothing available even as basic as a decent pair of shoes.

The Great Communicator

Looking at the empty shelves reminded me of one of the stories by US President Ronald Reagan.

Whatever you may think of Ronald Reagan—that he was the greatest US president in our lifetime or an easygoing Hollywood actor rumored to change into pajamas early in the evening so he could comfortably watch reruns of the *Carol Burnett Show* before going to bed—love him or hate him, few people deny that he was a successful president.

Stories abound that President Reagan could walk into a room when the tension was so thick that you could cut the air with a knife and he could immediately clear the air with some positive words. Reagan always carried the perfect knife—his ever-present sense of humor and his cheerful disposition. The Soviet Union was usually the butt of his relentless jokes about the communist system.

One of his famous stories centered on the chronic shortages in the old Soviet Union. It went as follows:

A man walked into a government-run car dealership.
He had finally accumulated enough rubles from working hard all his life to buy a car.
There was a long wait. He waited patiently.
The clerk eventually served him, asked him to fill out some lengthy forms, checked every detail for accuracy, stamped the form, and gave the man a piece of paper bearing the image of Vladimir Lenin's head and said:

"Here you go. Today is the 14th of the month.
Come back exactly three years from today and your vehicle should be ready.
And don't inquire in between because the answer will be the same. 'It is not ready.'
You really don't want to waste your time bothering us. It won't help things move faster anyway.'

"Great," the man replied as he nodded his head in agreement and prepared to leave.

Near the door, as an afterthought, he returned and asked the clerk:

"Morning or afternoon?"
The clerk could not believe his ears and said:

"Mister, we are talking about three years from today. What difference does it make right now whether morning or afternoon?"

"The plumber is coming in the afternoon," the man replied as a matter of fact, as he turned and left.

We are as free as the Americans
Another one of Reagan's jokes was about an American in Moscow's Red Square boasting to a Russian as to how free Americans are to express their views on anything or anybody.

"I can stand in front of the White House and say 'to hell with Ronald Reagan' and nothing will happen to me," the American proudly proclaimed.

"What is the big deal about that?" the Russian replied. "I, too, can stand in front of the Kremlin and say 'to hell with Ronald Reagan,' and nothing will happen to me either."

Where in the world is J.C. Penney?

Ronald Reagan's jokes about the scarcity of things were amply demonstrated as two of my friends and I entered the rickety store elevator to go to the third floor of the giant GUM store. Just before the elevator door closed, eight Romanian men stormed into the elevator uttering words that sounded more like expressions of frustration than jubilation.

I was wearing a pair of relatively new saddle oxford shoes that I had bought at an Atlanta shopping mall just prior to the Russia trip. I had not realized that of all the everyday necessities people needed in Russia, good shoes were the scarcest and the most prized possessions at that time. We learned during the trip that self-respecting, well-to-do Russians would kill (metaphorically speaking, that is) for a decent pair of shoes.

One of the Romanians in the elevator spotted my shoes. Pretty soon all eight men were staring at my feet as if I were a sacred Buddhist Monk that deserved to have my feet (with the saddle oxfords donned, of course) worshipped.

Enough staring, please!

Finally, the silence was broken. The Romanian closest to me happened to speak a little bit of English. He said: "Where? Where?"

I figured that he wanted to know where I had purchased the beautiful shoes. I could not remember where I had purchased them. The only American store name I could come up with under such lascivious yearning in a hurry was J.C. Penney.

"J.C. Penney," I said confidently.

The man closest to me, who seemed to be their leader, was the only person who heard me clearly. He turned to the man next to him and whispered "J.C. Penney." The second man turned to the third and said (with the same low decibel intensity), "J.C. Penney." Within a few seconds the J.C. Penney name had been conveyed methodically from one Romanian to the next, finally completing a full circle through the cavernous elevator car like fast-

269

falling dominoes.

The falling dominoes game was not over. The English-speaking leader inquired again, "Where? What city?"

"Atlanta, Georgia," I replied.

I then recalled our trip to Tbilisi, Georgia, and added, "America."

He shook his head in disappointment and softly announced, "Atlanta, Georgia, America."

A hush fell over the crowd. Disappointment was evident. J.C. Penney was in America, far beyond their reach. It was hopeless. Forget it. The saddle oxfords were not going home to Romania. For the Romanians it was back to the stores in Bucharest to look for decent shoes.

Chemical bonding

The time had come for our final dinner in Moscow before the long flight home.

Svetlana showed up early at our hotel. She seemed relaxed, although more subdued than usual. Dressed in a pink velvet outfit and with freshly styled hair, she looked like a different woman from the one who had greeted our rowdy group on our arrival in Moscow almost a month earlier. She was pretty. Gone was the harsh authoritative look. She smiled at me. The stone had been softened after persistent chipping and polishing, so much so that I was beginning to feel that we were forming a bond, a chemical bond.

I sensed some chemistry between us, but I was not sure. I had been there before when I thought that I had developed a bond with someone, but it would later prove to be a flimsy one, the kind that falls apart with any change in body pH. Nevertheless, this time the feeling seemed different. Being a scientist, I needed to know what kind of bond Svetlana and I had formed.

In my quest for clarity, I presented my chemical bonding hypothesis to the group earlier that afternoon at our daily vodka break in Professor Dietrich's room.

For many years there has been considerable debate among the scientific community on the relative order of the strength between various types of chemical bonds, including the ionic bond, the covalent bond (common to all organic compounds), the hydrogen bond, and the dipolar-dipolar bond.

I knew that the bond Svetlana and I had formed could not be the ionic or the covalent variety. Such bonding required much larger entropy (or shall I say investment in energy?) for formation. We did not experience combat together, although by reading through all of my Soviet stories and my earlier interactions with Svetlana, you might think otherwise. Maybe we did experience some combat, but it was more against each other than as allies. Hence, this seemed to be an unrealistic description for our relationship. Beyond that, to determine whether it was hydrogen or a dipolar-dipolar bond, I needed help from an expert in molecular physics. Luckily we had such an expert in our group. "How fortunate," I said to Hahn, the physicist and oilman among us: "Hahn, you may not believe it, but I think I have bonded with Svetlana. I need your help to decide what kind of chemical bond best describes this relationship, and please don't laugh."

"Jas, I don't want to deflate your ego, but like a lot of people from India, you seem to be a little overly sentimental in such matters. Being a scientist, I must rise above any sentimental crap and state the facts even if they deflate your ego."

"Okay, man. I can take it. Please give it to me straight." I was slightly irritated with his ethnic generalization.

"The bond you have with Svetlana is what is known as 'dipolar-dipolar,' sometimes referred to as a 'dipolar-dipolar attractive force.' This sounds nice, but this kind of bond is held together only by a thin thread. It can fall apart with just a loud whisper."

He continued, "Stronger ionic bonds require a hell of a lot more energy and are, therefore, more sustainable. All you have my friend is dipolar-dipolar, so don't get too excited," Hahn expertly declared.

"Okay, I am not heartbroken. I will take dipolar-dipolar, considering the little investment of energy I put into it," I replied.

Hahn continued, "Remember, a dipolar-dipolar bond is weak. I mean, it is like a cheap imported toy that falls apart at the very first use." He was trying to explain the complex physics to me in layman terms.

"I got it, man! I have a PhD in chemistry," I said in an irritated voice. "Do you think I have been pushing brooms up till now?"

Hahn smiled and said, "I am happy if you are happy, my man. Forget about physics or chemistry; just go for it."

Bonding confirmed

Proof of our bonding came soon enough on the bus the next morning. Svetlana motioned for me to sit next to her. *What a privilege!* I thought. I did not think Svetlana would ever willingly invite me to sit next to her. I must have been dreaming—a frequent occurrence as my friends and the readers of my stories have learned.

Sitting next to her during the bus ride to the airport, I tried to convince Svetlana that the group sincerely appreciated her patience and that I knew that almost all of us were a "royal pain in the bu—" My thoughts quickly reverted to another part of the anatomy than the one first intended and altered my sentence to, "Neck!"

"Go on, go on," she said. "I know what you were going to say. No need to feel embarrassed. Yes, you guys were a pain in the . . ." she hesitated, never completing the sentence.

"Go on and complete your sentence. I think I know what you mean," I told her.

She smiled and continued, "All in all, you guys were reasonably well-behaved. I had fun, although according to INTOURIST protocols, we are not supposed to acknowledge this. But I must confess to you that some nights I would go home and tell my cousin about the bullsh!t I heard all day from you Americans and we would burst out laughing. 'Crazy Americans!' we would say in unison." She laughed at her own story and ever so slightly squeezed my hand.

We were two minutes from the Moscow airport. I asked again if I could do anything for her after all that she had been through.

"Send me a postcard from Michigan," she said softly, her right eye showing the hint of a tear. "And if you ever come back to Russia, when times are different, I will let you read my palm."

Two weeks after returning home I mailed Svetlana a package, including a picture of Skaya, Svetlana, and me in the Moscow airport departure hall. I also enclosed a nice silk scarf as a token of my appreciation.

I never heard back from her.

My guess is that the photograph of Skaya, Svetlana, and me is gathering dust somewhere in the KGB's files in Moscow, and the fine silk scarf still adorns the neck of the wife of a retired KGB functionary.

I occasionally think of Svetlana and I hope that she is healthy and safe. Because that's what I do.

BIOGRAPHIES

Jas Singh, PhD, CIH

Jas Singh is a board-certified industrial hygienist whose implausible journey from a dusty village in Northern India to the pinnacle of the consulting profession is a story of inspiration, sadness, intrigue, joy, romance, and a fair bit of wonder. The chronicles of his social, travel, cultural, and work experiences are a remarkable reflection of life experienced through two extreme lenses—the poverty of rural India to the lavish boutiques of Beverly Hills, California, USA. Dr. Singh grew up in a humble farming family in a small village in Punjab, India. He excelled in school as a youth, was encouraged by his family to do his best in all his endeavors, worked hard, received an excellent education, built businesses, and has since shared his knowledge and experiences with many young people in his global professional network.

For the past thirty-five years, this globetrotting scientist has been sharing knowledge about occupational health, safety, and environmental protection throughout five continents. His sense of humor and humble origins allow him to bridge cultural divides with ease and gives him intimate access to people from different nationalities, ethnicities, social classes, and religious groups. Through his tales, chemistry becomes fun. Enjoyment of his life stories requires no background in physics, chemistry, or industrial hygiene—just an open mind and a mutual appreciation for experiencing cultures from the streets.

Jas now lives in a small—but not so dusty—village in Hawaii. He can be reached at: JSinghCIH@gmail.com.

Gregory Beckstrom

Greg is a geologist, writer, and business manager who has worked in energy exploration, environmental services, and engineering/consulting for the last thirty-five years. He has held a wide variety of positions, including exploration geologist, technical writer, marketing manager, business development manager, and operations manager. He has provided EHS consulting services to oil & gas companies, industrial manufacturers, mining companies, law firms, and a host of other private-sector enterprises located throughout the world. Greg and Jas used to work together at Golder Associates and developed a common bond based on their humble origins and mutual appreciation for exotic foods, dusty villages, hotel club lounges, and paths seldom traveled. Greg is now a vice president with AMEC Environment & Infrastructure.

He lives in Minneapolis, Minnesota, and can be reached at: greg.beckstrom@amec.com.

Carol Nagan

Carol is an artist who lives and works in Minneapolis, Minnesota. She was exposed to the professional side of art at an early age through her favorite uncle who was a commercial artist working for Gamble-Skogmo. She graduated from college with a fine art degree and spent many years working as a draftswoman, technical illustrator, and graphic designer. She is now doing what she loves—being an artist and illustrator. Illustrating this book has been a highlight of her career. She feels like she knows Jas because of these wonderful stories.

Carol can be reached at: carol@cd-design.org.

ABOUT THE ILLUSTRATIONS, MAPS, AND PHOTOGRAPHS

All original illustrations in this book were provided by Carol Nagan as a service to Jas Singh. They are the property of Jas Singh. All rights reserved.

The map of European Settlements in India in the Hotel Cidad de Damen story is a self-published work by Luis. Wiki map reference, University of Pennsylvania with reference verification. It is used with permission.

The photograph of the damaged buildings in the Great American Salvage Operation story was taken on January 17, 1994, and is courtesy of the FEMA Photo Library and is used with permission. Go to: www.fema. gov/photolibrary.

The photo of the Khajuraho Monument in the Sewage Chemistry for Dummies story was taken by Elizabeth Sonne, friend of Mary Singh's, and is from her personal collection. It is used with permission.

The photo of the Apsara painting in the ISO-Certified Brothel story was taken by Jas Singh and is from his personal collection. It is used with his permission.

The map of the Former Eastern bloc area border changes between 1938 and 1948 in the Behind the Iron Curtain Introduction is a file from Wikimedia Commons. Author is Mosedschurte, June 1, 2009. Permission is granted under the terms of the GNU Free Documentation License, Version 1.2 or later version published by the Free Software Foundation.

The maps of Armenia, Georgia, Ukraine, and Azerbaijan in the Behind the Iron Curtain stories were provided by John Moen, Managing Director, WorldAtlas. They are used with his permission. Go to: www.worldatlas.com.

The group photos in the Behind the Iron Curtain chapters are from Jas Singh's personal collection. They are used with his permission.

The photo of the GUM department store in the Romancing the Stone story is from Richard Seaman and his web site: www.richard-seaman.com. It is used with his permission.